REMODEL
OR MOVE?

REMODEL OR MOVE?

MAKE THE RIGHT DECISION

Dan Fritschen

ABCD PUBLISHING LLC
SUNNYVALE, CALIFORNIA

ABCD Publishing LLC
1030 East El Camino Real, #150
Sunnyvale, CA 94087
888-825-4169 www.abcdpublishing.com

Individual sales. This book is available through most bookstores, or can be ordered directly from remodelormove.com or ABCD Publishing at the address above.

Quantity sales. Special discounts are available on quantity purchases by corporations, associations, and others. For details contact the "Special Sales Department" at the address above, or by e-mail at specialsales@abcdpublishing .com.

Printed in the United States of America

Library of Congress Cataloging-in-Publication Data
Fritschen, Dan, 1964-
  Remodel or move? : make the right decision / by Dan Fritschen.
      p. cm.
  Includes bibliographical references and index.
  ISBN 1-933007-69-9 (pbk.)
  1. Dwellings—Remodeling—Cost effectiveness. 2. House buying—
Cost effectiveness. I. Title.
  TH4816.F75 2004
  643'.12—dc22   2004018330

Every attempt has been made to present accurate and timely information. We have used many sources, including our own professional and personal experiences, to compile the information found in this guide. The fields of real estate and remodeling are constantly changing and have substantial regional differences. These variations will account for many of the changes in references, resources, information, statistics, and techniques that will occur by the time that this book is purchased and read. Many individuals and professionals may have opinions that differ significantly from the information included in this book. Therefore, nothing contained in this book should be construed as an absolute with regard to this subject matter or be considered a substitute for real estate, legal, financial, construction, or building practices advice from licensed professionals and governmental agencies.

To Barbara

Contents

Introduction

TO REMODEL YOUR HOME or move to a different one is a question many homeowners face. In most cases, homeowners do not consider the question carefully enough. Congratulations to you for taking the time to research the options and ensure you are making the best decision.

This book is written because most of the six million people who move[1] and the one million who undertake a major remodel[2] and the seventy-four million who continue to find ways to improve their homes[3] each year do not have the information necessary to make the best decision. They may be happy with the decision they make, but making the wrong decision could cost them tens of thousands of dollars and many months of free time.

This book will take you through the decision process—from the reasons why people move or remodel to the costs associated with each choice. To enable you to be confident that you made the right decision, this book gives equal importance to both the financial and emotional considerations. Once you have an idea of the choices you will make, go online to www.remodelormove.com and complete the

Remodel-or-Move Calculator. It will calculate the cost for your remodel project and the cost for you to move. Then it will analyze your preferences regarding other important characteristics of your home. With this information, the Remodel-or-Move Calculator will provide a summary of the results and a recommendation to remodel or to move. Finally, after your decision has been made, this book will take you through the first steps of remodeling or moving.

Over the past twenty years, I have had the opportunity to buy and sell houses, design and build additions, and consult with others on the issues associated with moving and remodeling. In my experience, most homeowners think remodeling is too expensive, too complicated, and too risky. This is not to say that remodeling is always the right choice. If you want to have a bigger yard or live near better schools, moving is usually the right choice.

Remodeling can be so much more rewarding than simply pulling up roots and moving to a home that has all the room and features that you need. In addition, the true costs of moving and remodeling are not well understood, especially the cost of moving. Keep in mind that moving can cost a considerable amount of money, depending on the distance between your present home and future location. Some believe that a move is free since most of the expenses are taken directly out of the proceeds from the sale, not your wallet. However, it is still your money, and paying 10 to 15 percent of your home's value to move to another house needs to be considered carefully.

When you consider the time and inconvenience required to move or to remodel, the two options are surprisingly similar. To sell your home, you have to plan for the sale, decide where you want to move, find a new home, sell your existing home, pack, move, unpack, and settle in. It can take six to twelve months from the start of the process to the point where you can relax and call your new house home. Likewise, a remodel includes the planning stage, the construction phase, perhaps moving all or part of your possessions to storage, the move back in, and the recovery phase. This can also take six to twelve months.

No matter what type of person you are—analytical or emotional—the Remodel-or-Move Calculator will allow you to understand both the financial and emotional considerations before making your decision. Even if it is not an easy one, you will be well prepared to make the best decision possible.

Once you have made your decision, this book will provide you with the basic how-tos for remodeling and selling. With this information, you can decide if you need to do more research or if you are ready to start knocking down walls or scheduling open houses. If you need more information, Web sites, books, and professionals in your area can help with any questions you have. You will find recommendations in the appendix.

Before we begin, one statement must be made: your home is not just an investment. It is true you can make money selling your home, just as you can make money selling your time. It is also true that most of your net worth is probably tied up in the house, but, like your time, your home is worth so much more. You don't sell your time to the highest bidder. You want to do work that you enjoy as well as earn money. Your home is the same. You spend most of your time in your home, and you need to enjoy it regardless of cost. Having a south-facing window that lets the sun warm you on a cold morning, a bathtub that is surrounded by tile in your favorite color, a closet that holds all your possessions perfectly, or a workshop that fits all your tools can't be valued the way a money market or mutual fund account can. So, as you work to make the best decision, please do not ignore the quality-of-life or the financial issues. To ignore either will leave you poorer.

1

Ten Reasons to Move

To make the best remodel-or-move decision, it is important to understand all the reasons that you have to remodel and all the reasons that you have to move. This chapter reviews common reasons to move. As you read each reason, consider how much it applies to you and note how strongly you feel about it. Your reasons for moving and how strongly you feel about them will be used in the Remodel-or-Move Calculator.

The Size of Your Family Has Changed

Of the many reasons to move, the size of your current home is the most common. Many young couples purchase a cozy two- or three-bedroom 1,000-square-foot home that suits their situation perfectly. The home has a master bedroom, a guest bedroom, and possibly a home office. A single living area, with couches and an entertainment center, provides the couple with ample space for the two of them and their visitors. As they start a family, the first child moves into the guest bedroom and toys take over the living area. The perfect house for two becomes too small for three or more. Take the example of Tim and Leslie.

> Tim and Leslie bought their first home in an older part of town. The 1,000-square-foot, two-bedroom house was affordable and close to their jobs. At the time, the real estate market was booming so they thought they had better buy a house while they could. After a few years and several small remodeling projects, they began to consider their future. They intended to start a family in a few years, and their current home was not in the best school district. They lived on a street with a lot of traffic, which was fine for two adults but less desirable for children, who have a habit of chasing after balls regardless of where they roll. More importantly, the house was suitable for the two of them, but too small for a family. So, with their growing incomes and a need for a more family-friendly environment, Tim and Leslie found a 1,700-square-foot, three-bedroom home in the same city but in a better school district and on a quieter street.
>
> Now they have lived in the larger house for five years and grown to a family of four so they are laying plans for a remodel. They would like to update the kitchen, remove a few walls to open the floor plan, and add two more bedrooms and one bathroom. Construction is set to begin as soon as the youngest child begins kindergarten.

Later in life, after the children have grown, couples once again find their home to be the wrong size—in this case, too big. Four or five bedrooms and 3,000 square feet gives a family of four room to grow and enjoy some amount of privacy, but after all the children have moved out, many rooms are no longer used. If you find yourself in this situation, moving is typically the best solution.

You Want Better Schools

Unless your children attend private school, where you live usually dictates which school your children attend. This is great for building a sense of community in a neighborhood as children can walk or bike to school together. A school nearby can also cut down on travel time for dropping off and picking up your children.

The importance of schools can be seen in the difference in the price of similar homes, even within a few blocks of each other. It is not uncommon for homes to vary in price by tens or even hundreds of thousands of dollars based on the school attendance area that they are in. The example below describes how Sue and Larry dealt with this challenge.

> When Sue and Larry were shopping for their first home, they had a limited budget but wanted to live in a nice community. Without children, schools were less important to them than the price, condition, and location of their new home. They also reasoned that it would be eight to ten years before they would have school-age children. To try to buy a home now in a great school district might prove futile since the quality of schools and attendance areas could change in ten years. They decided to buy a comfortable home in a nice neighborhood and then move again when their children reached school age.
>
> Since the arrival of their children, they have added on to their home to give their small family more room.

Because their home was one of the smaller ones in the neighborhood, the addition made their home average in size, making it possible for them to recover the cost of the remodel when they do finally decide to move.

Your Commute Is Too Long

A common reason to move is the change of a job. Be it out of state or to another community, most people will commute only so far. If you want to have a shorter commute, then moving may be the only answer. The big decision to be made in this case is do you want to keep your present home and just rent in your new location? Renting out your current home and moving to a new home closer to your work can make a lot of sense unless you are confident that your move to the new area will last more than just a few years. The costs of selling your home (10 to 15 percent of its value) and buying a new home (another percentage of the new home's value) can add up to a lot of money. If you end up staying for only a year or two and then moving back, renting in the new location is a better option.

You Don't Like Remodeling

Since you are reading this book, you are probably at least considering remodeling your home. Remodeling is not for everyone. No matter how it is accomplished, two factors are unavoidable: the inconvenience and the decision making. The inconvenience can be as little as not using your kitchen for a day while it is being painted or as much as moving out for six months while some major work is done. For some, any inconvenience is too much so a move may be the better option.

Unless you turn your house over to an interior designer and an architect and give them total control, you will also have to make some decisions. The decisions are not simple ones. Unless you have reasonable spatial skills, deciding on what can be done with a room by moving walls, windows, and doors could prove to be a challenge. Even using pictures from magazines or books can help only so much, since changes will have to be made to accommodate your floor plan as well as to meet other requirements. If you don't like making decisions—if even deciding whether to remodel or move is unpleasant for you—moving may be the better choice.

Also, consider how your family will react to the expense, stress, and inconvenience of a remodel project. All family members should, at a minimum, understand and accept the inconveniences. Remodeling can be a big strain on a marriage, especially if one spouse is heavily involved in the remodeling project and the other spouse hates it. Take time early on to set expectations about what will happen during the remodeling process, what involvement each spouse will have, and what the schedule will be.

One good solution for managing the remodel process as a family team is to allow family members to make decisions about specific aspects of the projects. With hundreds of decisions to be made, there are plenty to go around. Try letting children take the responsibility for deciding on how their rooms will be decorated and where the furniture will be placed. It could be a fun and educational research project for them. They can look at possibilities at the library, in store showrooms, and on the Internet. Perhaps one spouse could concentrate more on the decorating of the rooms while the other focuses on the floor plan design and works with the contractors. Sharing the responsibility with all members of the family can make the job easier for everyone as well as avoid making family members feel left out. Remember to respect the decision of whoever has spent the most time working on the solution or has been assigned the task of making the decision. While others in the family may not agree with the decision, making a family member feel involved and respected is usually more important than which shade of green a room is painted or whether a window is 3 or 4 feet wide.

You Don't Like Your Neighborhood

Each neighborhood has it own characteristics. Some have big yards; some have small. Some have sidewalks and streetlights; some don't. Some have many rental homes; some have few. Some are full of 1,000-square-foot houses; some only have 3,000-square-foot houses. In some neighborhoods, children play on the street all day long and friendly neighbors stop by to chat every day. In others, people keep to themselves and rarely wave as they drive by each other on the way to work. As much as we all would like to change some features of our neighborhood, many are out of our control. If the neighborhood doesn't meet your needs, a move may be your only solution.

If you do decide to move to a new neighborhood, don't assume you can easily tell what the neighborhood will be like from a few visits and assurances from a real estate agent. It is very difficult to understand all the intricacies of a neighborhood without spending months or years there. Even though you may not love your current neighborhood, carefully consider your decision to move to another neighborhood you believe is better. Make sure that it truly is better.

There are several good ways to learn about a neighborhood. One tried and true way is to visit a few of the neighbors. Knock on the door and introduce yourself as a possible new neighbor. If you go to five houses and no one will talk to you, then it is likely that the neighbors won't be too friendly after you move in. If you end up visiting with a family for a half hour, then you probably will fit in very well. It is likely that you won't experience either of these extremes, but you definitely will have a better feel for the neighborhood after talking to your would-be neighbors. Also try to visit the neighborhood during different times of the day and on weekends. The streets may be quiet at three in the afternoon, but busy when commuters drive through the neighborhood as a short cut to avoid traffic.

You can also get statistics on the neighborhood, city, and schools from the resources found in the appendix of this book.

Your Home Has a Bad Floor Plan

You may not be able to remodel the house you live in. It may be too costly to make it what you want. If you want a kitchen in front and a family room facing south, but your home has the kitchen in back and a family room facing north, a remodel may be too expensive to be practical. The added cost of rearranging rooms is difficult to recover when you sell because those who are looking to buy your home may not have your need for specific room layouts. You are essentially adding significant expense for very personal aesthetic reasons.

Some homes, because of lot size, building codes, or physical barriers, may not lend themselves to remodeling the way you want. Building codes can limit the type and size of additions as well as their appearance. This is especially true when homes have views; for example, you may not be permitted to add a second story or any addition if it blocks your neighbor's view. Also, some private communities do not allow any exterior modifications, or if they do, all your neighbors must approve them.

If for some reason your house can't be remodeled, then moving will be your only option. When you look for your next home, remember to inquire about these restrictions and make sure that you like the floor plan. That way when it is time to consider a remodel, you will be free to consider all of your options.

You Don't Like Your Yard

For many, the yard is an integral part of a house. A yard is land to call your own, a place to plant a flower or vegetable garden, or an expanse of grass that you take pride in keeping green and manicured all summer long. No matter what it is about a yard, most of us want one. The question is, how big a yard do you want? There can be good reasons for wanting a smaller yard as well as a larger yard.

In urban areas and suburbs, yards are often small. On a 4,000- or 6,000-square-foot lot, there is little room if you want more than one of the common yard features: a lawn, a garden, a dog run, or a pool. Elaborate landscaping can be expensive and is difficult to justify for any reason besides your own enjoyment. Even if you choose to add every possible feature to your existing yard, the yard is only so big; you cannot add on to it. Therefore, if you want a yard with all the amenities and room to grow, it may be wiser to look for a new home.

If you are on the other end of the scale, with little interest in maintaining a yard, your choices are limited. You can let it go to weeds (but this won't make you popular in the neighborhood), or you can hire a gardener to keep your yard looking manicured. Finally, you can move to a new home with little or no yard. Many private communities feature maintenance-free homes. This is especially attractive to owners who may also be interested in downsizing from a four- or five-bedroom home to something smaller that is easier to care for inside and out.

Closely related to the need of many for a yard around their home is the need for privacy. Many houses in new subdivisions, especially those with two-story homes, lack privacy. The homes have as little as 10 feet of separation, and many windows face each other. Going outside offers no privacy because the second-story windows in adjacent homes offer ample opportunity for others to see you in your yard.

Your desire for a certain type of yard and a certain level of privacy are an important factor in determining whether you should move or remodel. Take the time to understand what you like and don't like about your current home as well as how important improving it is to you.

Remodeling Is Too Expensive

You can profit from an intelligent remodel, but in reality, some remodel projects are not worth the investment or effort. Ultimately, you should spend your money the way you would like, so if installing a $20,000 custom-built bathtub imported from Italy in your $80,000 house makes you happy, then go ahead and do it. However, many of us would like to make decisions that are practical and will bring us years of enjoyment. In these cases, we need to weigh the cost and benefit of a remodel versus a move. If your dream is to live in an affluent neighborhood with large lawns and 4,000-square-foot homes and your neighborhood is full of bungalows with single-car garages, the chance of your getting what you want for a reasonable amount of money by remodeling is unlikely. In addition, it is cheaper to build new than to remodel. Starting from scratch is easier than working with what is already there. So before you take the plunge on a major remodel, consider what you could get if you invested the money in a new home.

Your Home Is Already the Largest and Nicest in the Neighborhood

You may have remodeled your home before and now your home is the largest and nicest in the neighborhood. This is great if you love the neighborhood and plan to stay in the home for years to come. However, if you have been remodeling your home because it needs updating and you still need to add additional space or amenities, then a move may be a good idea financially. The houses near yours have a large influence on your home's price. If the homes next door are palaces and you are living in a cottage, your cottage will sell faster than if it was two blocks away and next door to other cottages. Before making the remodel-or-move decision, take a look at your house and your neighbors' houses. If you have the nicest or largest house in the neighborhood and need more space, moving is the better solution since your payback on additional improvements won't be very great.

You Will Likely Move in the Next Few Years

If there is a wrong time to remodel, it would be right before you move. Don't go through the expense and the inconvenience of remodeling and then put the house up for sale the next month or the next year. Moving right after a remodel can be costly unless you managed your remodel project very carefully to minimize the cost and maximize the market appeal of the work that was done. Also, moving takes away one of the biggest benefits of remodeling: enjoying the results. If there is a reasonable chance of your moving in the next twenty-four months, it may be better to move now to get the house you want rather than to remodel. An alternative is not to do anything and wait until you are assured of your housing status. That way, when you undertake a remodel project you will be confident of enjoying it for years to come.

If you are like most homeowners, you found that several of the reasons to move discussed here apply to you. You may even have additional reasons that were not included in this chapter. To make the best remodel-or-move decision, make a list of all your reasons to move. Next to each of your reasons, add a comment on how strongly you feel about it. To complete the Remodel-or-Move Calculator at www.remodelormove.com you will need to have your reasons to move and to remodel. Take a moment now to complete the table below or make a similar chart on a separate piece of paper.

Table 1.1 Reasons to move

Number	Reason	Does this reason apply?	How important is this issue?
EXAMPLE	The size of your family has changed.	☑ Yes ☐ No *We have 5 people living in a 2-bedroom house*	☑ Very ☐ Somewhat ☐ Unimportant
1	The size of your family has changed.	☐ Yes ☐ No	☐ Very ☐ Somewhat ☐ Unimportant
2	You want better schools.	☐ Yes ☐ No	☐ Very ☐ Somewhat ☐ Unimportant
3	You don't like your commute.	☐ Yes ☐ No	☐ Very ☐ Somewhat ☐ Unimportant
4	You don't like to remodel.	☐ Yes ☐ No	☐ Very ☐ Somewhat ☐ Unimportant
5	You don't like your neighborhood.	☐ Yes ☐ No	☐ Very ☐ Somewhat ☐ Unimportant
6	You don't like your home's floor plan.	☐ Yes ☐ No	☐ Very ☐ Somewhat ☐ Unimportant
7	You don't like your home's yard.	☐ Yes ☐ No	☐ Very ☐ Somewhat ☐ Unimportant
8	Remodeling is too expensive.	☐ Yes ☐ No	☐ Very ☐ Somewhat ☐ Unimportant
9	Your home is already the biggest/nicest in the neighborhood.	☐ Yes ☐ No	☐ Very ☐ Somewhat ☐ Unimportant
10	You will likely move in the next few years.	☐ Yes ☐ No	☐ Very ☐ Somewhat ☐ Unimportant

2

Eight Reasons
to Remodel

ow that you have considered all the reasons to move, you need to consider the reasons to remodel. Many of them are the opposite of the reasons to move. For example, if you like your neighborhood or the local schools, remodeling can be a great opportunity to keep what you have and add to it. As you read each of the reasons to remodel, consider how it applies to you and note how strongly you feel it about. Record your thoughts on the chart at the end of this chapter. Like your reasons for moving, your reasons for remodeling will be used in the Remodel-or-Move Calculator.

You Enjoy Remodeling

Remodeling, in a word, is fun—a lot of fun. Remodeling can be very rewarding and enjoyable, especially a year after the work is done. It may not be a lot of fun while it is in progress, but in total it can be a very satisfying experience. What makes all the pain, suffering, and expense fun? You have opportunities for self-expression and you have control.

If you are an artist, an architect, or a novelist, you can point to something and say, "I conceived and built this." The rest of us can get the same rewarding experience through remodeling our homes. You have total control over the project. You decide on whom to hire, when to start, what to do, what color it should be, and how it should be finished. The artist in you can run wild. Take for example Justin and Ashley, who took on the challenge of an older home that needed updating and were able to profit from and enjoy the process of the remodel.

> Justin and Ashley took on a remodel project to update a home that was mostly the original 1957 design. While they liked the basic floor plan, they wanted a new kitchen, a larger master bath, and an additional bedroom. Over three years, they drafted the designs, took part in some of the work, and now have an updated house with a great floor plan. While improving their house was their primary objective, the artistic and mechanical challenges were also major reasons they embarked on the project. They hired out most of the work to be done but still did a portion themselves to reduce the cost and, more importantly, because they enjoyed learning new skills. They enjoyed remodeling their current home so much that they will likely be moving in two years to a home in a better school district. They are looking for a "tear down," a home that is in the right location but is in poor shape or very small. This will enable them to demolish the house and build a new home in its place.

You'll Get Exactly the Home You Want

Another significant advantage of remodeling is you most likely will get a home that more closely fits your needs than if you move. If you move, you will have to envision yourself living in one of five to ten homes that best meet your requirements when you are house hunting. It is unlikely that any one is ideal. Each one will require you to make a variety of compromises, such as the neighborhood, the yard, the floor plan. The list goes on and on. And it is likely that you will want to make some changes to your new home as well. In the end, you may be trading old problems for new ones. Plus you will be paying for the privilege of moving.

If you remodel, you already have a list of features that you know, through years of experience, could be improved in your home. After you add the family room, remodel the kitchen, and complete all the other projects on your list, there is a good chance your remodeled home will better fit your needs than if you moved.

You Like the Neighborhood and Neighbors

Not all neighborhoods are created equal. Some have large mature trees, sidewalks, large front yards, and stately homes with well-kept yards. Other neighborhoods have few trees, cars parked in the front yards, and a highway behind the back fence. If you like the neighborhood you are in, then that is a powerful reason to stay. Many times your neighbors and neighborhood are hard to duplicate in a new house. The local park, the trees that spread across the street, the Fourth of July block party—all of these impact your quality of life significantly, and you can keep them if you remodel.

You Can Avoid Buyer's Remorse

Usually, a decision on which house to buy is made quickly, within hours or days. When the decision to buy a house is made, you have to make many compromises, including the location, size, yard, floor plan, and price. No home will meet all of your requirements perfectly. There are also many unknowns when you decide to buy a home. You may not know the neighbors, and the neighborhood may be unfamiliar. All of these changes in your life when you move are based on a decision made very quickly. That can lead to buyer's remorse—regretting the decision you made.

Remodeling a home is a lengthy process that requires months to start. Almost everyone who remodels his or her home is happy with the results. They may not have enjoyed the process or may not have been satisfied with the contractor, but the result is most often what they wanted. This is because someone who is remodeling a home is only changing the house. They are not changing the neighborhood, the yard, or the schools. Buyer's remorse is more common with a home purchase than a remodel because the decision is made much faster and more aspects of your life are changing.

Remodeling Can Be a Good Investment

If your home is medium to small for your neighborhood and is in need of improvements and if your additions are appropriate in style and quality for the house, then remodeling could be a good investment. If you manage the costs well, it is reasonable to assume remodeling will add more value to your home than the cost of the remodel. Take the example of Justin and Ashley, who invested $80,000 and will make a $40,000 profit on that investment when they sell their home.

Justin and Ashley had purchased their 1957 house three years earlier at a 10 percent discount because of the condition of the house. They liked the neighborhood and bought the least expensive house they could find. They began updating and remodeling as their finances improved. They made many cosmetic improvements, such as new maple kitchen cabinets to replace the plywood ones and new marble bathroom tile to replace the original pink tile. These changes, as well as the addition of a fourth bedroom, have made the house average in size for the neighborhood, a good place to be from an investment perspective.

To make these improvements, they invested $80,000 over three years and probably could sell the house today for $120,000 more than if they had done none of the improvements. In addition, if they sold today, with all the improvements, the house would likely sell faster and with fewer contingencies than if they had left it in its previous state. This "ease of sale" is an important value because selling your home is a stressful experience. Most of us value a fast, low-risk sale.

Remodeling can also be less expensive than moving. As an example, imagine you own a $200,000 home. The home was built in the 1970s and needs updating in every room, but generally you like the floor plan, the neighborhood, and the yard. Your options are to remodel and update your current home or to move to a new home that is selling for $300,000. To sell your home will cost between $20,000 and $40,000, as is shown in chapter 3.

What kind of remodeling could you do with the same amount of money? As examples in chapter 4 will show, you could paint the interior, install new interior doors with new hardware and add baseboards and crown molding, which when combined will make a big improvement in the home's appearance. After all these upgrades, you will still have money left over to do a minor update of the bathrooms and kitchen, which could include new knobs on the cabinets, painting the cabinets, and new faucets. All of this could be done for much less than the cost to move, saving you money, allowing you to keep the house that you love, and improving your quality of life.

You Like Your Yard

Many older homes are blessed with mature trees, a large lot, and years of landscaping efforts by you and the previous owners. If you are fortunate enough to live in a home with a yard that you love, then it may be difficult to find a home with a yard that you like as much. As you look at homes that you may be interested in buying, don't underestimate the cost and time required to take a yard that is no more than dirt and weeds and turn it into your dream garden. Large shade trees can take ten to twenty years to grow from seedlings. Large mature shrubs can take three to ten years to grow. If a yard is important to you and some of the houses that you could possibly move to don't offer anything close to what you have in your current home, a remodel may be the best solution.

You Like Your Location

You may now live in an older part of your town that is close to your work, around the corner from downtown, and has a view of the mountains in the distance. While you like the location, you have dreams of living in a new, larger house with high ceilings, an expansive entry, and a gourmet kitchen. A possible choice is to move to a new home that offers all of the styling and features you are looking for. But the new homes in your city are ten miles away in the middle of a cornfield that was turned into a subdivision a few weeks earlier. With no schools, shopping, or other features you like nearby, the decision is difficult. You like the new houses but not their location. In this case, you can have the house of your dreams and the location you love by remodeling.

You Like Your Home's Floor Plan

The floor plan, how the rooms are positioned in a home, has a big impact on how you can use the rooms and how you feel about the home. If your current home has a floor plan that you like—a large kitchen with morning sun, an out-of-the-way spot where you can read a book without being interrupted, appropriate-size bedrooms (small ones for the children or a den and a big one for you), a perfect spot for a Christmas tree or a grand piano—it may be hard to re-place these features when you move to a new house.

Take a moment now to consider the reasons you may wish to remodel, including ones that are not mentioned in this chapter. Then list all of your reasons on a sheet of paper or the table below and mark any that are very impor-tant to you, ones that have a major impact on the quality of your life. These could be your yard, if you have spent ten years landscaping it, or your neighbors, if they are some of your best friends. These reasons to remodel and their importance to you will be used in the Remodel-or-Move Calculator.

Table 2.1 Reasons to remodel

Number	Reason	Does this reason apply?	How important is this issue?
EXAMPLE	You like remodeling.	☑ Yes ☐ No *I have always wanted to build my own home*	☑ Very ☐ Somewhat ☐ Unimportant
1	You like remodeling.	☐ Yes ☐ No	☐ Very ☐ Somewhat ☐ Unimportant
2	You will get exactly the home you want.	☐ Yes ☐ No	☐ Very ☐ Somewhat ☐ Unimportant
3	You like the neighbors and neighborhood.	☐ Yes ☐ No	☐ Very ☐ Somewhat ☐ Unimportant
4	You can avoid buyer's remorse.	☐ Yes ☐ No	☐ Very ☐ Somewhat ☐ Unimportant
5	Remodeling can be a good investment.	☐ Yes ☐ No	☐ Very ☐ Somewhat ☐ Unimportant
6	You like your yard.	☐ Yes ☐ No	☐ Very ☐ Somewhat ☐ Unimportant
7	You like the location.	☐ Yes ☐ No	☐ Very ☐ Somewhat ☐ Unimportant
8	You like your home's floor plan.	☐ Yes ☐ No	☐ Very ☐ Somewhat ☐ Unimportant

3

How Much Does It Cost to Move?

The move option is often considered easier than a remodel, but sometimes it can be more difficult and expensive. This is due to the difficulty and expense of selling your home and the time and effort required to find a new home. The costs of a typical move include preparing your home for sale, hiring a real estate agent, closing the sale of your home, purchasing your new home, moving, and, finally, decorating and furnishing your new home. Like the chapter on remodeling, in this chapter you will be provided with a description of a variety of expenses and, using the Remodel-or-Move Calculator in chapter 6, you can choose which expenses will likely apply.

The following table describes some of the costs to move and their typical amounts. Each individual's circumstances are different so these costs may or may not apply to you, and the actual cost could be higher or lower. In any case, you should be familiar with these costs as you decide whether to remodel or move.

Table 3.1 Costs of moving

Description	Cost
Real estate agent commission	6 percent of sales price
Home preparation prior to selling	Zero to $9,000
Moving	$500 to $5,000
Furniture, appliances, and decorating your new home	$1,000 to $10,000
Miscellaneous fees and services	2 to 3 percent of home sale price
Increase in property tax	0 to $10,000 per year
Total for a $200,000 home	$15,000 to $50,000+

Getting Your Home Ready to Sell

To sell your home fast and for the highest price, you can take many steps to make it appealing and inviting to prospective buyers. You will have to decide what your home needs and what it doesn't. Also ask your real estate agent, if you hire one, about dressing up your home for a faster sale.

On the outside of your home, consider the following improvements to sell your home faster and for a higher price.

- Your landscaping and the lawn needs to be in good shape. If you don't have time, hire a gardener. Invest in annual flowers and other plants to brighten up the place. Low-voltage lighting can also add charm for a minimum cost, and most buyers will look at your home at night before making an offer. Total cost: $1,000.

- The mailbox, street number, and exterior lighting should all be clean and functional at a minimum, but upgrading these items will improve the appearance of your home. Total cost: $250.

- The front door and exterior of the house should be painted if they look weathered. Total cost: up to $3,000.

- Gutters and the roof have a big impact on the appearance of your home, but unfortunately, at $10,000 to $15,000, these expensive updates won't provide much of a payback if you replace them in order to sell your home.

- Fences should be in good repair and painted/stained if needed. If they are in poor condition, consider replacing them. Total cost: up to $750.

- Total cost for outside preparation: $4,000.

On the inside of your home, consider the following improvements to make your home more appealing to potential buyers.

- Minimize family photos and religious or cultural items on display. You want the people viewing your home to imagine that the home is theirs as soon as they walk in the door. All of your personal items will just remind them it is someone else's house. You will need to pack these items away when you move anyway, so do it sooner and sell your home faster and for a higher price.

    Other places to start your packing early are closets, attic, basement, and garage. A home with a few, neat items stored in these areas invites buyers to be impressed with the large amount of available storage and envision how they could use all this space. Full storage areas don't appear very large and don't help would-be buyers imagine how they could use the space. These areas are often the easiest to pack and move anyway. Often, they are full of boxes still packed from your last move, seasonal clothes, or decorations. Pack these items and move them out! Unless you have access to free storage, assume $500 for storage costs to cover the period from listing your home to moving these items into your new home.

- Visible cracks and damage in the walls, windows, floors, and other surfaces should be repaired. Potential buyers will typically spend only a few minutes walking through your home. They want to put it quickly on the list of possible homes they may purchase or ones they are not interested in. Seeing a few cracks here and there in the drywall or windows will quickly eliminate your home from consideration for many buyers. Only buyers looking for a fixer-upper will be interested in your home, and they won't offer you top dollar. Total cost: up to $1,000, depending on the extent of any problems.

- Windows and doors should all operate smoothly and be easy to open and close. All the locks should be working. An inexpensive way to dress up your home is to change the interior doors and doorknobs if the existing ones are damaged or outdated. In most areas, for as little as $200, you can purchase a new door cut to match your existing door. Spend another $20 for a lock set, and you have a brand new stylish door. A typical three-bedroom home has about ten to twelve of these doors, for a total cost of $200 for doorknobs or $2,000 for new doors and doorknobs.

- Clean, clean, clean. Windows, walls, floors, kitchen, bathrooms, everything should be clean all the time.

You want prospective buyers to have access to your home whenever they want, so assume that would-be buyers will be dropping by anytime between 8 a.m. and 8 p.m. Keep the rooms picked up and clean; dust and vacuum regularly. (You may like the way your house looks when it's always organized and clean enough to keep up this tidiness after you move!) You may wish to hire a professional cleaning service to do a thorough cleaning initially. Total cost: up to $150.

- Have all appliances, heating and cooling systems, plumbing, and light fixtures in good working order. If one light switch doesn't turn on a light, one ceiling fan doesn't work, or one sink leaks, buyers will imagine ten other things that need to be repaired. They will decline to make an offer or lower the offer by $3,000 just to reserve enough cash to fix the items they discover. Invest $1,000 and have everything in good working order.

- Bathrooms should have no smells, no mildew, and no discolorations on any fixture (tub, shower, sink, or toilet). Toilet seats and faucets are inexpensive and fast to replace. Chips in the porcelain can also be repaired. Tile should have the grout cleaned, repaired, if needed, and recaulked. If your home has smaller bathrooms with a tub/shower combination remove the shower curtain and keep some nice props such as fancy soaps, a jar of bath beads, and a matching towel set in the bathtub. Total cost: up to $500 per bathroom, depending on the condition.

- Total cost for preparing your home on the inside: up to $5,150.

Real Estate Agent's Commission

The majority of homes are sold using a real estate agent to
market the home and represent the homeowners in the
negotiation and sales process. Real estate agents charge
from 3 to 6 percent of the home's selling price for their services. The following are benefits of using a real estate agent.
This subject is covered in more detail in chapter 7, "So You
Have Decided to Move—What's Next?"

Setting the Sale Price

A real estate agent who is experienced and successful selling homes in your area can use his or her knowledge to
help you select a sale price. The agent's experience and
pricing strategy could help your house sell faster and potentially for more money.

Staging

To sell your house fast and for the most money, the appearance of your home is very important. An agent will look
through your home and offer advice on how to improve its
appearance.

Marketing

A real estate agent has ready access to and experience in
the use of several valuable marketing tools including the
Multiple Listing Service (MLS), print advertisements in
newspapers and magazines, Internet listings, and a network
of other real estate agents.

Negotiating

An experienced real estate agent can offer expertise in negotiating the best deal for you.

Closing

Many details and paperwork are required to complete the sale of a home. An experienced agent can manage all of these, allowing you to work on preparing to move.

If you choose to sell your home without the help of a real estate agent, you may save the cost of the listing commission, typically 3 percent of the home price, but you may pay a commission (also 3 percent) to the buyer's agent unless you find the buyer yourself. Many of the other costs will be the same, such as the closing and home staging costs. The costs that may be higher if you sell the home yourself are marketing and the cost to hire an attorney with real estate contract expertise. When using an agent, these costs are typically covered by the 6-percent listing commission. Without an agent, you should assume you will spend several hundred dollars at a minimum to market your home and another several hundred dollars, at a minimum, to consult with a real estate attorney to create, review, and complete the home sale documentation.

Inspections

A presale home inspection allows you to get an independent appraisal of your home's condition before you put it on the market. This is very helpful since it is likely that the buyers of your home will require a home inspection as part of the contingencies of their offer. By having an inspection done yourself, you won't be surprised by what the buyer's inspector finds. If you choose not to have your home inspected prior to listing it and a buyer's inspection turns up some needed repairs, you will be under pressure to make a fast decision. Should you make the repairs and allow the deal to go through, drop your price and let the buyer make the repairs, or not make the repairs and let the buyer possibly rescind his or her offer? Unless you are prepared to make these types of decisions quickly, it is much easier to have an inspection before you put the house on the market and make the necessary improvements. The cost of a home inspection is typically a few hundred dollars.

Should You Remodel Before You Sell?

Some improvements can help a home sell for more money and should be considered by homeowners, especially when updating is appropriate. The rooms that have the largest impact on a buyer's opinion of a home are the front entry, bathrooms, and kitchen. An updated kitchen with new appliances, countertops, and attractive cabinets will make buyers excited and can make the difference when they are deciding between two homes. But the cost and time necessary to make such kitchen improvements isn't worthwhile if you are preparing to sell. Instead, focus on making only minor, cosmetic repairs and improvements such as

- Painting the interior

- Painting stained or dated cabinets and woodwork

- Adding new door hardware

- Replacing faucets and fixtures in the bathrooms and kitchen

- Replacing light fixtures, switch plates, and electrical outlets and covers

- Replacing baseboard and door trim

A bathroom can take on a whole new appearance with a few small changes. In the following example, a bathroom was changed from an eyesore to a comfortable and clean area by installing new light fixtures, new mirrors, and paint to complement the tile colors.

> When Cheryl and Steve moved into their home, the primary bathroom had peeling wallpaper and a fifteen-year-old vanity and sink. Seeing this room as a first priority, Cheryl and Steve removed the wallpaper, painted the bathroom, and replaced the sink and vanity. For $250 and a Saturday's effort, they took an eyesore and made it look clean and fresh. Had the owners done this prior to putting the house on the market, they most certainly would have recovered their expenses and would have sold the home much faster.

Don't expect a buyer to have the imagination to be able to see what your home will look like if the carpet was clean, windows and fixtures were not broken, and the walls were freshly painted. Take the time and spend the money to clean, repair, or replace these items before you put your home on the market. As you consider these improvements, it is important to select the most generally appealing colors and shades. Your intention is to make the house look clean, in good repair, and appealing to the most people possible. Consider white or off-white paint for the walls and trim and beige for the carpets. These generic colors are neutral and won't distract from the house's other features. They will help to sell the home faster and for more money.

Moving

What does it cost to move? It may cost as little as a few hundred dollars to rent a truck and buy the necessary packing materials. A typical three-bedroom home will take about 40 to 80 hours to pack and about half that time to unpack, for a total of 60 to 120 hours. You can do this yourself, if you have time. You may choose to hire a moving service to assist with the packing or just with loading and unloading the truck and moving the larger pieces of furniture. For estimation purposes, use $25 per hour if you want to hire a full-service moving company. For packing and unpacking, the total would be between $1,500 and $3,000 for labor plus the cost of travel time and the truck. These transportation costs can add as little as $500 for a local move (less than 100 miles) or as much as $5,000 or more for a long-distance move. All the above estimates are based on a typical three-bedroom home with an average amount of belongings; your cost for moving may vary.

Buying Your New Home

Once you have sold your home and packed up your possessions, you are only half done with your move. The following is a list of common expenses and tasks that homeowners face when they are looking for and then moving into a new home:

- Look at houses on the market and decide how much you can afford and which house you want.

- Make an offer and negotiate the price.

- Get financing.

- Set up new gas, electrical, water, telephone, cable, insurance, and other services.

- Send change-of-address information to friends, relatives, banks, and others.

Frequently, you will want to make improvements to your newly purchased home before you move in. This could be as simple as repainting and laying new carpets or as extensive as remodeling a bathroom or kitchen. It is often more convenient to make these changes before you move in to avoid any added inconvenience.

Furniture and Appliances

Most moves involve the purchase of new appliances or furniture. This is often done since moving is a great time to update your furniture, to add new furniture for your larger home, or to replace appliances that you sold with your old home. While these can be major expenses, in the thousands and tens of thousands of dollars, they are similar to the costs you are likely to incur if you remodel your existing home.

Inspections

After finding a new home that suits your needs, you will want to have the home inspected by a qualified home inspector to determine if the home has any unseen structural or other problems. Inspectors will check the condition of the home from top to bottom and make notes on any problems or potential problems that they see. This will include the condition of the home's heating system, central air-conditioning system, interior plumbing, and electrical systems; the roof, attic, and visible insulation; walls, ceil-ings, floors, windows, and doors; the foundation, basement, and visible structures. If possible, be present for the inspec-tion and take time to discuss the results with the inspector as well as to review the inspection report when it arrives a few days later.

Closing Costs

"Closing costs" is a catchall term for the miscellaneous costs that are paid by the buyer at the time the title to the home is transferred. These costs may include points on the new mortgage, title charges and insurance, a credit report fee, a document preparation fee, a mortgage insurance premium, inspections, appraisals, prepayments for property taxes, a deed-recording fee, homeowner's insurance. The amount and type of charges will vary from state to state.

Increase in Property Tax

In many parts of the country, property taxes make up a large percentage of the cost of owning a home. Since your property tax can vary from 0 to 4 percent, it is important to consider the impact of moving on your property tax bill. In California, as an example, the value of homes for property tax assessment changes at a maximum of 2 percent a year as long as you own your home. When you buy a home, the value is reset at the purchase price. Therefore, if you owned your home for many years and it has appreciated during that time, your property taxes will likely go up if you move.

For example, assume the current assessed value of your home is $100,000 and the home you are looking at costs $400,000. Your property tax rate is 1.25 percent so you are currently paying $1,250 a year. When you move to the new home, your property tax bill will be $400,000 times 1.25 percent, or $5,000 per year, a significant change from the $1,250 you have been paying. Assuming you would have lived in your original home for an additional ten years if you had remodeled, you could have saved $42,500 in property taxes over those ten years if you had decided not to move.

What it will cost you to move is dependent on many factors, including whether you use a real estate agent and the condition of your home. Considering all of the possible expenses in this chapter can give you an estimate of the cost, but the true costs won't be known until after you have moved since you are likely to have many unforeseen expenses. As you weigh your options to move or remodel, make sure that you do not underestimate the cost and time required to move.

4

How Much Does It Cost to Remodel?

I f you are weighing the decision to move or remodel, chances are that the cost of the remodel will be a major factor in your decision. This chapter will provide you with a description of the choices you will need to make before you can get an estimate from a contractor or from the Remodel-or-Move Calculator at www.remodelormove .com. The decisions you will need to make are

- Will you hire a general contractor or manage the project yourself?

- Will you hire a designer/architect or do the design work yourself?

- How much of the construction work will you do? None of it? Some of it? A lot of it?

The decisions you make on how the remodel project will be managed and who will do the work can have a significant effect on the cost. If you choose to hire someone to manage the entire project and you decide to let professionals handle everything during the project, then you can easily spend twice as much as a homeowner who gets a little dirty, tolerates some of the inconveniences that are a part of a remodeling project, and does some of the work. If you have no interest in doing any of the remodeling work yourself, then you can skip this chapter until after you have made your remodel-or-move decision. If you choose to move, then there is no need to come back to this chapter. However, even if you choose to remodel using a full-service contractor, it will be worthwhile for you to read this chapter to understand some of the construction terms and the process.

The Major Costs for Remodeling

Each of the cost categories below will be explained briefly so that you can make a decision about how you are most likely to proceed. You will need to note how many of the projects you would like to do and how many you want to hire out in order to complete the Remodel-or-Move Calculator. Specifically, you will need to decide if you want to hire a designer or do the design yourself, if you want to hire a general contractor or manage the project yourself, and if you want to do any of the construction work.

The following is a list of the primary tasks required in medium-to-large remodel projects. To estimate the cost of your remodel project and to determine if you want to remodel at all, it is helpful to understand these terms, how each task impacts the remodeling process, and if there are any tasks that you would want to do yourself.

- Project management

- Design (architecture and engineering, if necessary)

- Permits (issued by the city or county after approval of the project)

- Site preparation and demolition

- Foundation work (usually poured concrete walls or slab)

- Framing (the structure for the floor, walls, ceiling, and roof)

- Electrical

- Plumbing

- Heating and cooling

- Interior and exterior surfaces (drywall inside, siding and shingles outside)

- Finish flooring (hardwood, carpet, tile)

- Doors and windows

- Cabinets, fixtures, appliances

- Finish work (trim, baseboards, painting)

- Your first open house (drinks, food, entertainment)

Project Management

Project management is the role of coordinating the different phases of a project. All the work could be done by a single person, which makes project management easier, but more likely it will take a variety of different skilled tradespeople. If different people are needed to complete the project, one person needs to coordinate when they come and go, answer questions that arise on the job, and generally make sure the work is done right. The project manager also ensures that all the permits are secured, that inspections by the city and county are arranged, and so on.

There are three different options for you to decide on for project management:

- Hire a general contractor.

- Manage the project yourself.

- Hire a fee-based project manager.

General contractors will do the project management for you and usually do some of the work themselves. Often, the general contractor will do the framing, doors, windows, finish work, and fixtures and then subcontract with others to do the rest of the work. However, there are no hard and fast rules here. General contractors typically will provide a fixed bid for your project but may also charge for the hours they work and the materials they use (time and materials). General contractors will not manage the project for free. They will include in their fixed bid their hourly rate of $50 to $100 per hour of management activities and/or mark up what the subcontractors charge them by 10 to 200 percent. They will also take care of paying all the subcontractors so instead of writing twenty checks, you only have to write one, to the general contractor. You don't have to decide now! Moreover, when you use the Remodel-or-Move Calculator try entering different project management options to get a feel for the difference in total project costs.

Another option is to manage the project yourself. As described above, the role is primarily coordination and decision making. Even if you hire someone else to do the project management, you still have to make many of the decisions, so taking care of the coordination is a very reasonable task. Even if you don't have construction experience, there are resources available (see appendix) to help along the way. Don't let a lack of experience sway you too much. The more important question is, do you have the time and interest to take on and complete this project? It can take an hour or two each day to keep up, make calls, and make decisions. If you have other things you would rather do, go do them and hire a general contractor. If you don't enjoy learning new things nor the stress that comes along with hiring subcontractors, again, you are better off hiring a general contractor. However, if you have good people skills, strong organizational skills, and the time, and you enjoy learning new things, it can be quite an adventure.

A big part of taking this on yourself is that you will have to find and hire subcontractors. With a big project, this may take time. If you consider that you may need six or more different skills (electrical, plumbing, drywall and texture, and so on), you will have to interview quite a few people before picking the right person or company for each task. This is where networking helps. One good subcontractor will often know others. Ask your neighbors and friends; look in the newspapers. Good subcontractors are out there! Chapter 8 has suggestions on where to find contractors and subcontractors.

The benefits of being the project manager yourself include the pride of knowing you had a hand in the creation, the empowerment that comes with knowing that you made the decisions along the way that produced fine results, and the money you will save. As described above, without a general contractor a project can be 20 to 50 percent less expensive. This savings can allow you to make more improvements or keep more of your money in savings.

A third option is a compromise between the two. Most homeowners choose to use a full-service general contractor, many act as their own general contractor, and only a few choose a fee-based project manager. Fee-based project managers will do all the project coordination and subcontractor-hiring tasks at an hourly rate, $50 to $100 per hour. They do not mark up the subcontractor charges; you will usually pay them directly. This can be a very attractive option because it could save you money and take away a lot of the burden managing the project. Unfortunately, these fee-based individuals are not nearly as common as general contractors, and it may be hard to find one in your area that meets your requirements.

Design

Once you have decided whether you will manage the remodel project yourself or hire a general contractor, you now have to decide how to design the project. In order to obtain the permits from your local government and ensure that the remodel is completed to meet your needs, you will need a good set of design drawings. For uncomplicated projects, like installing new fixtures in a bathroom, finishing a basement, or installing new windows, these can be as simple as sketches on notebook paper with dimensions and notes on materials. Detailed design drawings with full engineering calculations can require ten to twenty large complex drawings.

Both simple and complex remodel projects need complete designs to ensure good communication between you and the people doing the work. The project will run smoother and faster and turn out better if these designs are complete. You can imagine what could happen if you don't have a good design. Cabinets might not line up with sinks, electrical outlets could be in the wrong place, or windows could be too low or too high. The room for error is very large. However, complete designs don't have to be expensive or fancy. They just need to contain enough detail so that both you and whoever is doing the work understand what to do and how.

For simpler projects, like a bathroom remodel or window installation, you or a contractor can successfully create these design drawings. If you choose to do it yourself, you should spend time considering the design requirements of each specific project. Several resources are available (see the appendix) that can provide you with direction and design options. Most include sample designs to give you an idea of what your designs should look like. Study these, come up with the requirements for your project, document the designs as suggested and shown in these books, and review them with whoever is doing the work. Ask for their suggestions and comments. Include their ideas in a revision, and you should be ready to go.

For larger projects you will need more complex drawings in order to obtain construction permits. They will ensure the project is built to meet codes, as well as to meet your design goals. These larger projects include moving walls inside the house, building additions, and adding a second story. For these projects, you also have the option of hiring an architect. An architect typically has a university degree in architecture and is registered with the state government.

Of course, you can still do the designs yourself. In some cases, this is preferable because for many people creating the design is one of the most enjoyable parts of the project. Doing it yourself can ensure that you are not making unnecessary compromises on the design. In addition, you can save the cost of an architect. In order to do the designs yourself, you will need to do the same research as with a smaller project. Again, there are resources available in the appendix that provide design ideas as well as requirements. A simple $50 home design computer program will help keep your designs neat and legible as well as speed up the revision process. You can use the drawings that you do yourself for obtaining permits and instructing the construction workers, or you can contract with a draftsman or engineer to convert them into more traditional architectural drawings with necessary details. The cost of the conversion is minimal compared to having an architect create the design from start to finish since most of the architect's fees are for developing the initial designs and incorporating revisions, not for the final drawing process.

Regardless of how you prepare the designs, you or your architect/designer may have to hire a licensed engineer, if required by your local government, during the permit process to ensure structural integrity of the design, heating, ventilation, and air-conditioning (HVAC) requirements, soil reports, and so on. Like other professionals in the construction trade, rates for an engineer's services can vary greatly. Recently licensed graduates or retired/semiretired, part-time, or out-of-work engineers may be willing to do the work for a small home remodel job for a few hundred dollars.

To find engineers that can do the work for below-average cost, ask the contractors you interview if they have a recommendation. Also ask the permitting office since they see many engineers each day. Check the newspaper and online. Use a search engine and enter the type of engineer you are looking for as well as the area you live in. If you hire an engineering firm that primarily works on commercial properties, the same work could cost thousands of dollars. The design cost can range from very little to as much as 10 percent of the total project cost.

Remodel Payback

Don't assume all additions are the same when it comes to resale value. It is important to realize that the addition is for your benefit, but one day your home will be sold to someone else. Making your remodel and additions suitable for a wide range of potential owners will bring you both the ease of a quick sale and the highest resale value. I like to look at new tract homes selling in the area to understand what design features are currently popular. While you shouldn't take new tract homes as the height of architectural design, they do show you what the homebuilders consider to be the most salable features. Incorporating some of the features of these homes in your remodel design will help you achieve a home design that both makes you happy and has the widest appeal for other homebuyers.

Some changes don't have wide appeal and therefore won't provide much return on your remodel investment. One example is converting a garage into living space. The additional living space will be a plus for resale value, but the loss of the garage will have a negative impact because many prospective buyers will deduct the cost of changing the space back to a garage from their valuation of the home. This is a good example of what you can learn from new tract homes. Two-car garages are standard, and three are common in larger homes, so a house with no garage is definitely a drawback.

A remodel project to avoid is major exterior changes without an addition of square footage or other interior amenities. We have all seen the one colonial-style home in a neighborhood of ranch-style homes. The owners of the colonial had a desire to build the house of their dreams and may have accomplished that during their remodel, but it is unlikely that they will find someone that shares that specific idea of a dream house. So the tens, even hundreds, of thousands they spent on changing the facade without much change on the inside will likely result in a slow sale and little payback of their investment.

Another remodel project to avoid is one that substantially changes a portion of the house so that it doesn't match the rest. In the following example, homeowners who dreamt of a perfect kitchen but installed it in a less-than-perfect house will have a difficult time recovering the money they invested when it is time to sell.

> Ray and Beverly wanted to remodel their kitchen. The entire corner of the 1960s 1,500-square-foot ranch-style house was removed and replaced with a stylish modern kitchen—black countertops, custom cherry cabinets, a food preparation island, a breakfast nook, and a built-in 30-inch television. They had hired an expensive kitchen design contractor to do the design and the work, and the result was beautiful. However, the rest of the house, including the adjacent dining room and living room, remained in the original 1960s ranch style with flat hollow doors, dated carpet, and dark interiors. In this case, it is unlikely that someone will pay Ray and Beverly even 50 percent of what they invested.

Ray and Beverly invested all of their remodeling money on a single room instead of investing some money in updating other parts. For a few thousand dollars taken from their kitchen budget, the trim and door styles used in the remodeled kitchen could have been used throughout the house, and they could have installed a few strategically placed skylights and new floor coverings in the main living area. For the same cost, these changes, along with the kitchen re-model, would have made the entire home a nicer place to live and assured a greater return on the couple's remodel-ing investment.

The following table lists some of the remodel projects and the return they provide to the owner. Remember that these are all averages. You can expect a much better return if you use average quality materials and reduce the labor and general contractor's markup by doing some of the work yourself. You can also expect to achieve a much lower return if you hire a full-service designer and a contractor and use extravagant materials or choose designs that appeal only to you.

Table 4.1 Remodel project payback

Project	National average	Sample project payback
Major kitchen remodel	$43,804	75 percent
Bathroom addition	$15,519	95 percent
Master suite addition	$70,760	76 percent
Family room addition	$53,983	81 percent
Bathroom remodel	$10,088	89 percent
Window replacement	$ 9,568	85 percent
Basement remodel	$43,865	79 percent

(Reprinted with permission from the November 2003 issue of *Remodeling* magazine. Copyright 2003 by Hanley Wood.)

Permits

Once you have your design completed, it's time to get your permits. You can visit the planning and development office in your area or have your contractor do it for you as part of the service. The permit process varies widely, depending on location, and is run by your city or county government. You should call the planning office and describe your project to find out if you need a permit. The permit process requires that the renovation conform to state and federal requirements. Inspectors from your local government may visit the project at key points to ensure that basic workmanship standards are met. When you eventually go to sell your home, prospective buyers and lending institutions may want proof that alterations complied with local codes. Without a permit and inspection on record you have no proof. The homeowner must then apply for a permit with no guarantee that the remodel will meet the codes and face the possibility that the remodel must be redone or removed. This is costly and frustrating and could cause delays in refinancing or selling the home. Permits range in cost from tens to hundreds, even thousands, of dollars.

Site Preparation and Demolition

Site preparation is very important for larger remodeling projects because you need to protect the portions of your home and yard that are not supposed to be affected. Equipment and materials will be hauled in and out of your house so make sure you have an agreement with your general contractor or subcontractors as to when and where this can occur. Also make sure that you or someone you hire builds barricades around landscaping that you want to protect. Be extra careful to protect not only the trunk but also the roots of trees from damage.

On the inside of your home, seal all pathways, such as doorways, shared closets, and bathrooms between the areas that are being remodeled and the rest of the home. Closing the door between these rooms will not be enough to keep dust and noise out. You should seal the openings with plastic sheeting and good-quality masking tape.

If you have irreplaceable items, they should be stored far out of the way or, better yet, put in storage until after the work is done. Dust during construction projects manages to get everywhere in spite of how well you seal the construction area. For items that are seldom used and/or are hard to clean, the best solution is to put them into storage as well.

Some site preparation tasks you have to do yourself, and others you can hire someone else to do. The cost to hire someone is not too expensive, nor will you save very much if you do it yourself, but it is very much a project that you, the homeowner, can undertake if you choose.

Demolition is great fun for some people and unbelievable drudgery for others. This can involve tearing up floors, knocking down walls, and making many trips to the dump. Like site preparation, demolition is a part of the project most homeowners can do, but again, the cost savings will be minimal. Typical costs range from 1 to 5 percent of the total project cost.

Foundation Work

Changes to your foundation or construction of a new one will likely be required if your remodel plans call for the addition of rooms outside the current floor plan of the house. If your remodel plans involve new cabinets, flooring, and so on in current rooms, then your project may not require any foundation work. Unless you are experienced with the construction of foundations, this is not a good project for inexperienced homeowners. Foundation work can be up to 10 percent of total project cost.

Framing

Framing is required when walls, ceiling, or roofs will be
added or moved. If you are remodeling an existing bathroom
or a kitchen, then there may be minimal framing work. If
you are planning an addition, then framing will be a big
part of the project cost. Framing is generally fast work that
requires lifting and some strength. It can be a good project
for a homeowner, but professionals use many techniques to
save time and improve the quality of the end result that
most homeowners won't be familiar with. Only take this on
yourself if you have the time and determination to learn
what is needed to do a good job. Framing can account for
up to 20 percent of the total project cost.

Electrical

Electrical work is required in most remodels. It can be as
simple as changing a light fixture or as complicated as
adding circuits to the main electrical panels. The simpler
electrical work is a good project for homeowners who take
the time to learn about the proper techniques and safety
considerations. Working with the electrical panels, adding
circuit breakers, and so on should be left to a licensed electrician. Electrical work can account for up to 15 percent of
the total project cost.

Plumbing

If your remodeling plans include changes or additions to
your bathroom or kitchen, then some plumbing work may
be required. Like some electrical work, some simpler
plumbing projects are good tasks for homeowners. Also like
electrical work, you will need to take the time to learn the

skills required, but the cost savings can be significant. Plumbing work can account for up to 10 percent of the total project cost.

Heating and Cooling

Adding or moving heating or cooling ducts is often part of a remodel project. Simple ductwork is a good project for homeowners. More complex heating or cooling projects, such as installing or moving furnaces or air conditioners are much more difficult and should only be attempted by homeowners with lots of preparation. If you don't have the time or the interest in learning everything you need to know, then hire a professional to do the installation. Heating and cooling work can account for up to 10 percent of the total project cost.

Interior and Exterior Surfaces

Most remodeling projects require installation of drywall on the inside walls and possibly installation of siding or other material on the outside walls. Both can be done by homeowners, but some of the finishing of drywall and some types of exterior surfaces such as stucco involve skills that are difficult to master. Since these tasks are usually not too expensive (5 to 10 percent of the total project cost), you probably should have a contractor do it.

Finish Flooring

A homeowner can successfully install flooring. The savings in labor-intensive installations such as tile can be significant. Carpeting requires less labor to install so there is less to save by doing it yourself. If you have the time and interest, the more labor-intensive installations, such as tile, are good projects to undertake. You can save a significant amount of money, but since the finish flooring will be seen for years to come, make sure you use the right tools and take the time to learn how to do the job right. Flooring can account for 5 to 10 percent of the total project cost.

Finish Work and Fixtures

Installation of cabinets, fixtures, appliances, trim, base-board, and painting are all good projects for a homeowner to tackle. Since the cost of having these done by a contrac-tor can be up to 50 percent of the cost of a project, they make good candidates. An eye for detail, patience, and fol-lowing instructions are key to doing a good job with finish work. The finish work can be a major cost of a remodel, ranging from 5 to 50 percent of the total project cost.

The cost of the finish work and fixtures can vary a huge amount depending on the quality and type of products that you select. Kitchen cabinets can range from a few thousand dollars for low-end, white prefabricated units to tens of thousands of dollars for custom-built hardwood cabinets with complex trim and all the options. Likewise, the appli-ances in a kitchen can range from hundreds to thousands of dollars apiece depending on the manufacturer, the design, or the style. The variation in costs continues for the windows, bathroom and kitchen fixtures, doors, baseboard, and all the other finish work. As you plan your budget and begin selecting the products and designs that you will want for your remodel, carefully consider your goals. If your goal is to have a gourmet kitchen, then the price will be signifi-cantly more than a standard kitchen of the same size because of the expense of the appliances and other features.

To get an estimate of the cost of your remodel project using the Remodel-or-Move Calculator at www.remodelormove .com, you will need to decide the level of quality of the finish work and fixtures. As you can see from the sections above, the fixtures and finish work can represent 50 percent or more of some remodels, so pay close attention to how you answer this question in the Calculator to ensure your estimate reflects your remodeling goals. Your options for answers to this question in the Calculator will be economy, average, or expensive finish work and fixtures. Each type is described below to help you make your decision.

Economy

Economy finish work and fixtures means the use of the lowest-cost items. Examples include standard size, white kitchen cabinets built of particle board instead of hardwood; the lowest-cost bathroom sink, tub, and fixtures; aluminum windows or low-cost vinyl windows; flat hollow doors and minimal trim throughout the house; and the lowest-cost appliances and lighting fixtures. All of these products can give you years of service, but you will find that their useful life is much shorter than more expensive products. In addition, the wear and tear of use will be evident within the first few years as compared to more expensive products, which can be used for many years and still look as good as new.

Average

Average-quality finish work and fixtures include common hardwood kitchen and bath cabinets that are prefabricated and have standard features; a bathroom sink, tub, and fixtures that are commonly used; better-quality vinyl windows; molded doors in a wide variety of styles; and average-quality fixtures and trim used throughout the house. Average-quality products should hold up well for many years.

Expensive

Expensive-quality finish work and fixtures include hardwood, custom-built kitchen and bath cabinets with extensive additional trim and features; wood or vinyl-clad high-end windows; solid, natural-wood doors; high-priced bathroom, kitchen, and lighting fixtures throughout the house; and elaborate trim including wainscoting, solid wood, special-order baseboard, and crown molding. Expensive-quality products oftentimes will provide years and years of service, potentially outlasting the rest of the house.

Two Typical Projects with Sample Costs

The following two examples will help illustrate how different homeowners approached their remodeling projects. The first homeowners did a significant amount of work themselves. The second homeowners did not want to do any of the work themselves, but they did find that they could save money by hiring several contractors to do the work instead of a single contractor.

> Justin and Ashley's 1,500-square-foot ranch home built in the 1950s had an original second bathroom with pink tile extending halfway up the walls and around the tub and shower. The house also had original interior doors with many dents, original doorknobs with a dull finish, and dated baseboards and trim. The oak hardwood had been covered by carpet

since the 1970s. The floor plan was laid out well and the home met all of Justin and Ashley's needs, but it definitely needed to be fixed up.

Justin and Ashley decided a good place to start would be to replace the doors and trim, refinish the floors, and remodel the main bathroom. These steps, combined with new wall paint, would make the house look and feel brighter, cleaner, and not as dated.

The first step was to replace the doors. Taking the doors off the hinges, Justin and Ashley took them to a home improvement store that, for $200 each, cut new doors to fit the old frames. With hinges and door-knobs, each new door cost $250. The trim around the doors could be easily replaced. With lightweight wood filler and caulk, they filled the gaps and nail holes, and the doors and trim were ready for painting.

Next, they decided to refinish the floor in the hall-way, two bedrooms, and the living room, a total of 600 square feet. After getting several quotes, they selected a small contractor that specialized in hardware floor refinishing and installation. For $3 a square foot, $1,800 total, the floors were sanded and three coats of polyurethane applied. Justin and Ashley also had the floor refinishers install new baseboards at the same time the floors were refinished.

Justin and Ashley decided to replace all the fixtures in the bathroom. Also, since the walls were tiled, the drywall would need to be replaced. Justin and Ashley decided to tackle this job themselves since they were getting $10,000 quotes from contractors. Over two weekends, they removed the old fixtures and drywall and installed a new bathtub, drywall walls, new tile around the bathtub, a new vanity, countertop, toilet, light fixture and new vinyl flooring. Finally, it was time to paint. Again, Justin and Ashley got quotes from several painting companies before deciding on a small family-run company.

After one month of chaos and busy weekends, half of their home was fully updated and now looks great.

Below is a table describing the project and two prices, one for the way Justin and Ashley did it, with an investment of their time, as well as a typical estimate if a general contractor had done the work. Note that the prices listed below are from midpriced contractors in a high-priced area, Northern California. Contractor's prices can vary by 75 percent, depending on the type of contractor, your specific project, and the labor rates where your home is located.

Table 4.2 Justin and Ashley's remodel price comparison

	Some of the work done by the homeowners as described above	Hours of work	Quote from a general contractor
5 new doors with door trim, hardware installed	$250 each, $1,250 total	18	$400 each, $2,000 total
Hardwood floor refinished	$1,800	0	$3,000
Bathroom remodel	$2,000	50	$8,500
Interior painting	$1,500	0	$2,500
Total	**$6,500**		**$16,000**

The following example is of a much larger project where the homeowners wanted to hire others to do all the work.

Sam and Stacy own a 3,000-square-foot split-level house built in the 1970s. While the home is in good condition, they would like to enlarge and update the kitchen and add a breakfast nook by building a 200-square-foot addition into their large backyard. The project will require a complete addition including the construction of a foundation, floors, walls, ceiling, and roof to match the existing roofline. Electrical and plumbing work will be required as well as installation of the new cabinets, appliances, and flooring. A reputable contractor active in their neighborhood gave them a quote for $160,000, which included the design, permits, all materials, and labor for the entire project.

Sam and Stacy decided to get another bid and, after asking advice from friends who have done remodeling projects, chose to split the project into three different phases. The first will be the design phase. Working with a successful architect will allow them to better assess their needs and utilize an expert's advice on the kitchen layout, window and door placement, and other important details. In the second phase a general contractor will build the room addition, including the foundation, walls, roof, subfloor, and textured, primed drywall. The third phase will be the installation of the tile and hardwood floors, cabinets, and appliances.

The benefits of splitting the project in these phases are flexibility and potential cost savings. The flexibility comes from hiring an independent architect or designer, which allows you to find one that best meets your needs. While many contractors definitely can provide very good design services, hiring an architect yourself allows you to pick someone who will work with you. You will also have a wider range of contractors you can work with since not all offer design services. In addition, if you find you don't like your architect, you know that your involvement with that person will end when construction starts. If your contractor does the design work and things get off to a rocky start, you will have to stop the project and find another contractor or continue to work with the designer/contractor for the rest of the project.

The potential cost savings come from avoiding the fees that a general, full-service contractor adds to the cost of subcontractors and materials. General contractors can provide all the materials and labor for a project, but it is likely they will hire others to provide the design, cabinet construction, and flooring. The parts of the project they have subcontractors do are dependent on the general contractors' skill and capabilities. When general contractors hire subcontractors or buy materials, they add a fee to cover their costs. These fees can range from 10 percent to more than 100 percent and are included in the total price they charge. By hiring subcontractors yourself, it is possible to avoid these fees.

The table below outlines the cost with and without a general contractor for Sam and Stacy's remodel and shows the potential cost savings. By managing three subcontractors and shopping aggressively for good prices, Sam and Stacy were able to cut the cost of their remodel project in half, for a total savings of $85,000.

As you have seen in this chapter, the cost of a remodel can vary from a few thousand dollars to hundreds of thousands of dollars for a variety of reasons. The three major influences on the cost of your remodel are the scope of the project, the way you choose to manage the remodel project, and the quality of materials that you choose to use. By understanding each of these three influences, you can pay only for the services and materials you choose and ensure that you get the most for your money.

Table 4.3 Sam and Stacy's remodel price comparison

	Sam and Stacy manage the project themselves, hiring individual contractors for each phase	General full-service contractor
Architect	$10,000	Included
Foundation, exterior enclosure, drywall installation and texture, no finish work	$30,000	Included
Cabinets, finish work, and appliances with installation	$30,000	Included
Hardwood and tile flooring	$ 7,000	Included
Total	**$77,000**	**$160,000**

5

The Decision

Previous chapters have given you the most common reasons to move and to remodel and outlined the cost of both choices. This chapter prepares you to complete the Remodel-or-Move Calculator, which will aid you in making your decision. The Remodel-or-Move Calculator references the issues discussed in the previous sections and asks you to input your preferences. (Access the Calculator at www.remodelormove.com. You can also request a worksheet at www.remodelormove.com or by calling 888-825-4169.) With your preferences loaded, the Remodel-or-Move Calculator will automatically give you both a financial analysis of your options as well as a weighted result based on the emotional considerations.

How Do People Make Decisions?

We all make decisions differently. Some people use their "gut feeling," paying little attention to the advice of others or the financial details. Others rely mainly on opinions of others; they either ask many people and then follow the consensus or find someone they consider an expert and listen to his or her advice. Still others seek out numbers to calculate the best decision. They use warranty details, price, and other relevant facts they can find to aid in the decision. Whatever your decision method is, it is important to understand the process you use and its benefits and limitations.

The gut feeling method is to decide based on your emotions. Most of us have done this in situations such as playing the lottery, selecting what to order at a restaurant, or buying items in the grocery store. We don't give these decisions much consideration. They are safe decisions because we face minimal consequence for making a wrong decision. Consider selecting a steak for dinner versus pasta. Most of the time we order because we want a specific meal or taste. We don't analyze the calories, cost, or other considerations to make the decision.

Your emotions need to be a part of your decision to remodel or move. For example, do you like the neighborhood and do you like the yard? Some may like a specific neighborhood; others will not. It is a gut feeling decision.

The "other people's opinions" method delegates the decision to whomever you ask for assistance. For many decisions, such as which movie to see or which book to read, this method is fine. For the remodel-or-move decision, you should use this method with great caution. Many people you consult will not be experienced enough to give you an informed opinion or may have a vested interest in your decision. Only after several major remodeling projects will someone have enough perspective on the process to be able to recommend whether a remodeling project is a good idea for others. Few people with this amount of experience are skilled in providing beneficial consultations on whether you should remodel or move.

Experts in the field that you may consult often have a stake in your choice. For example, real estate agents and contractors stand to gain, depending on your decision. To garner an unbiased recommendation from them will be difficult. Also, being an expert in this area does not mean someone is good at helping you make a decision, only that he or she is able to sell your house or remodel it. Finally, no one can determine what is best for you. This is a life-changing decision that you should make using all available tools, including your gut feelings, your calculations, and the advice that you receive from others.

A "numbers" approach to the remodel-or-move decision is the basis for the Remodel-or-Move Calculator. It collects information about your specific situation and calculates the cost and the benefit of remodeling and moving. With this tool, even if you aren't a mathematical genius, you can easily get a calculated result, which will ideally agree with your gut feeling as well as the advice you receive.

Is Your Decision Simple or Complicated?

Now that you have considered in previous chapters the reasons to move and what it will cost and the reasons to remodel and what it will cost, you should determine whether you are a candidate for a quick decision. What is a quick decision? A quick remodel-or-move decision can be made by those who are staying in their present home for only a short period or by those whose remodel plans are simple and primarily cosmetic.

The length of time you will stay in your home is important because medium to large remodel projects take several months to plan and execute. If you plan to stay in your home for only one or two years, do you really want to spend some of that time remodeling, when you will be able to enjoy the results for only a short time?

If your remodeling plans are simple and your home is in need of the updates or if the changes will make it more attractive to buyers, you should most likely proceed with your remodeling plans. These types of projects include painting your kitchen cabinets, installing new doors, new windows, new crown molding or baseboards, and replacing the toilets and sinks in the bathrooms. These projects all are relatively low in cost and typically require just a few days of work. They also share their appeal to most potential buyers. So as long as your home needs these types of improvements and you shop around for the best price, you can proceed with confidence that you can recover most of your investment. Since the work will take only a few days, the inconvenience should be minimal, and you will have time to enjoy the changes before you move.

The rules that apply to a major remodel project also apply to smaller remodeling projects. The first rule is to be an assertive customer—shop around, get multiple quotes from different types of suppliers (for example, from a window store, an independent contractor, a home improvement store, and so on), and pay a fair price. The second rule is to make improvements equally across the entire house. Freshly painted cabinets and new pull knobs in the kitchen when the floor and countertop are in bad shape are not a good investment. Likewise, new bathroom fixtures when the bathroom door has big dents in it and needs painting are not a good investment either. Following the steps in the remodel section, make a list of these smaller improvements you would like to make and then determine how much you want to spend. Then spread the money around among the different projects so that all rooms in your house look consistently updated and follow a consistent style theme.

Regardless of how long you intend to live in a home, routine maintenance should be continued. Fixing appliances, repairing windows, painting inside and out, and fixing plumbing or electrical problems should be done regardless of how long you plan to stay in your home.

The following table provides a matrix of the types of improvements and how long you should stay in your home to justify the cost and inconvenience of each kind.

Table 5.1 Quick-remodel decisions

How much longer will you stay in the house?

Type of project	0 to 1 years	1 to 3 years
Install new doors and new windows	Okay	Okay
Paint or refinish bathroom and kitchen cabinets and install new hardware	Okay	Okay
Install new baseboards, crown molding, etc.	Okay	Okay
Lay new carpet	Okay	Okay
Refinish hardwood flooring	Okay	Okay
Install new hardwood or tile flooring	If necessary	If necessary
Remodel kitchen extensively	If necessary	If necessary
Add on a new bathroom	Not recommended	Okay
Finish a basement or attic	Not recommended	Okay
Add on a new room	Not recommended	Not recommended
Enlarge an existing living room or family room	Not recommended	Not recommended
Add on a new master suite	Not recommended	Not recommended
Add on a second story	Not recommended	Not recommended

Okay: Note that this recommendation assumes that you will use economy or average-grade material and spend an average amount on labor and the materials. If you are going to move in a few years, it is not a good idea to do even a simple remodel using expensive custom materials since you will have only a few years to enjoy the results and the next owners probably won't pay you even 50 percent of what you invested in these improvements.

Not recommended: These projects are not recommended if you will stay in your home for a short period since they are relatively expensive and/or can take several months to complete. Since you will stay in the home for only a short period after the remodel project, you won't be able to enjoy the results, and unless you manage the project very carefully to minimize cost you could easily spend more on the remodel than the additional amount you will realize from the improvements when you sell and move.

If necessary: If the items that you want to replace (for example, carpet, doors) are in very poor condition and will detract from your home's otherwise good overall condition when you sell, then you can consider replacing them. Before deciding to proceed with this improvement, make sure that you will enjoy the new items during the remainder of your stay in this house. Also use economy or average-priced materials and common designs and colors that will appeal to the largest number of buyers. Pay a reasonable amount for the installation.

At this point you should be equipped to make your remodel-or-move decision. In chapters 1 and 2 you made notes on the reasons that you want to move and the reasons you want to remodel. In chapters 3 and 4 you learned about the costs of moving and remodeling and made decisions about the role that you want to play in each process. In chapter 5 you learned about different decision-making styles and determined the best way for you to make your decision. Armed with this knowledge, you can move on to chapter 6, which will give you examples of completed Remodel-or-Move Calculators and results. With these examples, you can complete the Calculator online or use the *Remodel or Move Workbook.*

6

The Remodel-or-Move Calculator

The Calculator is key to collecting all of your thoughts and research on the question "Should I remodel or move?" The Calculator will consider your emotions, such as your feelings about the neighborhood, schools, floor plan, and other intangible factors related to your decision. The Calculator will also request that you enter the specifics on your remodeling project so that an estimate of the cost to remodel and move can be calculated. With this information, the Calculator will provide a summary of what it will cost for you to move, what it will cost for you to remodel, and finally, a specific recommendation on whether you should move or remodel and why.

In this chapter are three examples of the Remodel-or-Move Calculator, how they can be completed, and their results. Once you have reviewed them, go to www.remodelormove .com to access the online Calculator or call 888-825-4169 to request the most up-to-date Remodel-or-Move Calculator Worksheet. One will be sent to you by e-mail or mail. The online Calculator and Calculator Worksheets may differ from the example below because the team at www .remodelor move.com is always learning from homeowners

like yourself and regularly improving the Calculator to ensure you receive the best advice possible.

Remodel-or-Move Calculator Examples: A Major Remodel-or-Move Decision

In this example (table 6.1), Justin and Ashley like their
neighborhood but their house is outdated and too small, so
they are considering a major remodel. To make their current
home better fit their needs, they would like to add a master
suite, remodel the main bathroom and the kitchen, and
enlarge the family room. The four considerations for them
are, what will it cost to move to another house instead of
remodel? What would it cost to remodel? How much of the
work do they want to do themselves? And do they like their
current home enough to go through the remodeling process?

- They have seen houses that sell for $525,000 that
  have the extra space that they need as well as a
  similar yard and neighborhood to their current
  home's. Houses in their neighborhood similar to
  theirs have sold for $400,000. So a larger house
  that meets their needs would cost them the expense
  of a larger mortgage ($525,000 versus $400,000)
  plus the expense of moving.

- They answered in the Remodel-or-Move Calculator
  that if they remodel they will hire a general contractor,
  do some of the work themselves, and select average-
  quality products when they can. In this instance,
  that would mean commonly used windows and doors,
  standard height or raised ceilings, and a mix of
  carpet, common hardwoods, and tile for flooring.
  Using the information, the Remodel-or-Move

**Table 6.1 Remodel-or-Move Calculator—
Major remodel or move example**

Question	Answer
What could you sell your home for today?	$400,000
If you move to a house today that has all the features you need, what would it cost?	$525,000
What state do you live in?	California
How long have you lived in your home?	4 years
If you remodel your home, how much longer will you live there?	20 years
Who will do the basic design sketches and ideas?	An architect
Will you prepare the design drawings and apply for the permit yourself or hire someone?	Hire someone
Will you manage the project yourself or will you hire a full-service contractor?	Hire a general contractor
How much of the work will you do yourself (electrical, framing, painting)?	A little
What level of quality do you want in the finish work? Economy, lower-cost bathroom and kitchen fixtures, average-cost bathroom and kitchen fixtures, or custom and expensive bathroom and kitchen fixtures?	Average
Will bathrooms be remodeled? If so, how many?	Yes, 1
Will bathrooms be added? If so, how many?	Yes, 1

Table 6.1, *continued*

Question	Answer
Will the kitchen be remodeled?	Yes
If you are remodeling the kitchen, will the size of the kitchen change?	No
Will a second story be added?	No
How many rooms will be added, excluding bathrooms, living rooms, and family rooms?	1, a master bedroom
Will a living room be added or enlarged?	No
Will a family room be added or enlarged?	Enlarged
Remodel cost estimate, low range	**$90,000**
Remodel cost estimate, high range	**$145,000**
Estimated project payback	**$80,000**

Table 6.1, *continued*

Sell your house and move calculator

Question	Answer
Will you sell your home yourself or hire a full-service real estate agent?	Hire a real estate agent
Cosmetically, how would you rate your home today? Is it ready to sell or do you need to do some minor cleanup, repair, painting, etc.?	Good condition. It needs a little paint and other things fixed before selling.
Number of bedrooms in your home	4
Number of other rooms furnished or used for storage (family room, dining room, den, living room, basement, attic, sun room, recreation room, garage, basement)	7
Number of people living in the home	4
Number of years you have lived in the home	5
Will you hire a full-service mover or pack yourself and rent a truck?	Pack myself and rent a truck
Move cost estimate	**$55,000**
Expense of higher mortgage payments for the new house	**$60,000**

Table 6.1, *continued*

"Gut feeling" analysis

Question	Answer
Are you satisfied with your local schools?	Very satisfied
How important is the quality of the schools to you?	Very important
How do you feel about the distance from your home to your place of employment and other places you visit regularly (grocery store, etc.)?	Very satisfied
What are your feelings about the remodeling process? Are you excited about it or dreading it?	Excited
How important is the neighborhood to you?	Very important
Do you like your current neighborhood?	It is okay
Do you like the floor plan in your current home?	It is okay
Is the size and condition of your home's yard important to you?	Somewhat important
Do you like the size and condition of your home's yard?	It is okay
Is your goal to live in your dream house, or will any comfortable house meet your needs?	Be comfortable in my home

Table 6.1, *continued*

Remodel-or-Move Calculator Results

Question	*Answer*
Estimated move cost	$55,000
Expense of higher mortgage payments for the new home	$60,000
Total cost to move	**$115,000**
Remodel cost estimate, low range	$90,000
Remodel cost estimate, high range	$145,000
Project payback	$80,000
Remodel cost above payback	$10,000 to $65,000
Expense of higher mortgage payments for remodeled home	$26,000
Total cost to remodel	**$36,000 to $91,000**
Is it more expensive to move or remodel?	More expensive to move

> It could cost up to $60,000 more to move. Justin and Ashley's gut feeling score indicates a preference to remodel. The Remodel-or-Move Calculator recommends remodeling.

Calculator estimates that the cost of their addition would be between $90,000 and $145,000 minus the likely payback of $80,000 plus the cost of financing the $80,000.

• Justin and Ashley indicated in the Calculator that they like most aspects of their home, the schools, the neighborhood, and the yard.

Using the information that Justin and Ashley provided, the Remodel-or-Move Calculator determined the cost to move, approximately $96,000, and the cost to remodel, between $36,000 and $91,000. Since a remodel will likely be less expensive and Justin and Ashley's gut feeling score indicates a preference for remodeling, the Remodel-or-Move Calculator recommends that they remodel instead of move.

As with all recommendations by the Remodel-or-Move Calculator, it is very important to get quotes and advice from both real estate and construction professionals to determine the actual costs before making your final decision.

Example of a Minor Remodel-or-Move Decision

In this example (table 6.2), Janet and Stephen have a small family that has outgrown their home in Mississippi. To make their current home better fit their needs, they would like to add a family room. The four considerations for them are, what will it cost to move to another house instead of remodel? What would it cost to remodel? How much of the work do they want to do themselves, and do they like their current home enough to go through the remodeling process?

- They have seen houses that sell for $145,000 that have the extra space that they need, have a larger yard, and are in a better neighborhood. Houses in their neighborhood similar to theirs have sold for $110,000. So a larger house that meets their needs would cost them the expense of a larger mortgage ($145,000 versus $110,000) plus the expense of moving.

- They answered in the Remodel-or-Move Calculator that if they remodel, they will hire a general contractor, do none of the work themselves, and select economy products when they can. In this instance, that would mean economy windows and doors, standard ceiling heights, and carpet for the floor. Using this information, the Remodel-or-Move Calculator estimates that the cost of their addition would be between $23,000 and $37,000 minus the likely payback of $19,000 plus the cost of financing the $19,000.

- Janet and Stephen indicated in the Calculator that they like their home's location, are satisfied with the schools, the neighborhood, and the remodeling process but don't like their home's yard.

Using the information that Janet and Stephen provided, the Remodel-or-Move Calculator determined the cost to move, approximately $21,000, and the cost to remodel, between $4,000 and $18,000. A remodel will likely be less expensive, but Janet and Stephen's gut feeling score indicates a preference for moving. The Remodel-or-Move Calculator recommends that Janet and Stephen consider moving to a new home that has the yard that better meets their needs.

**Table 6.2 Remodel-or-Move Calculator—
Family room addition example**

Question	Answer
What could you sell your home for today?	$110,000
If you move to a house today that has all the features you need, what would it cost?	$145,000
What state do you live in?	Mississippi
How long have you lived in your home?	10 years
If you remodel your home, how much longer will you live there?	10 years
Who will do the basic design sketches and ideas?	An architect
Will you prepare the design drawings and apply for the permit yourself or hire someone?	My general contractor
Will you manage the project yourself or will you hire a full-service contractor?	Hire a general contractor
How much of the work will you do yourself (electrical, framing, painting)?	None
What level of quality do you want in the finish work? Economy, lower-cost bathroom and kitchen fixtures, average-cost bathroom and kitchen fixtures, or custom and expensive bathroom and kitchen fixtures?	Economy
Will bathrooms be remodeled? If so, how many?	No
Will bathrooms be added? If so, how many?	No
Will the kitchen be remodeled?	No

Table 6.2, *continued*

Question	Answer
If you are remodeling the kitchen, will the size of the kitchen change?	No
Will a second story be added?	No
How many rooms will be added, excluding bathrooms, living rooms, and family rooms?	None
Will a living room be added or enlarged?	No
Will a family room be added or enlarged?	Added
Remodel cost estimate, low range	**$23,000**
Remodel cost estimate, high range	**$37,000**
Estimated project payback	**$19,000**

Table 6.2, *continued*

Sell your house and move calculator

Question	Answer
Will you sell your home yourself or hire a full-service real estate agent?	Sell it myself
Cosmetically, how would you rate your home today? Is it ready to sell or do you need to do some minor cleanup, repair, painting, etc.?	Good condition. It needs a little paint and other things fixed before selling.
Number of bedrooms in your home	3
Number of other rooms furnished or used for storage (family room, dining room, den, living room, basement, attic, sun room, recreation room, garage, basement)	4
Number of people living in the home	4
Number of years you have lived in the home	10
Will you hire a full-service mover or pack yourself and rent a truck?	Pack myself and rent a truck
Move cost estimate	**$4,000**
Expense of higher mortgage payments for the new house	**$17,000**

Table 6.2, *continued*

"Gut feeling" analysis

Question	Answer
Are you satisfied with your local schools?	Somewhat satisfied
How important is the quality of the schools to you?	Very important
How do you feel about the distance from your home to your place of employment and other places you visit regularly (grocery store, etc.)?	Very satisfied
What are your feelings about the remodeling process? Are you excited about it or dreading it?	Indifferent
How important is the neighborhood to you?	Somewhat important
Do you like your current neighborhood?	It is okay
Do you like the floor plan in your current home?	It is okay
Is the size and condition of your home's yard important to you?	Somewhat important
Do you like the size and condition of your home's yard?	No
Is your goal to live in your dream house, or will any comfortable house meet in your needs?	Be comfortable in my home

Table 6.2, *continued*

Remodel-or-Move Calculator Results

Question	Answer
Estimated move cost	$4,000
Expense of higher mortgage payments for the new home	$17,000
Total cost to move	**$21,000**
Remodel cost estimate, low range	$23,000
Remodel cost estimate, high range	$37,000
Project payback	$19,000
Remodel cost above payback	$4,000 to $18,000
Expense of higher mortgage payments for remodeled home	$7,000
Total cost to remodel	**$11,000 to $25,000**
Is it more expensive to move or remodel?	Costs are approximately equal

> The costs to remodel and move are about the same. Janet and Stephen's gut feeling score indicates a preference to move. The Remodel-or-Move Calculator recommends moving.

As with all recommendations by the Remodel-or-Move Calculator, it is very important to get actual quotes from both real estate and construction professionals to determine the actual costs before making your final decision.

Example of a Major Remodel-or-Move Decision

In this example (table 6.3), Kurt and Marie have a growing family. Their home is too small and they are considering adding a second story. Their neighborhood is satisfactory, the street they live on is busier than they would like, and some of the houses nearby aren't kept as well as theirs. Their local schools are average, but they would like to live in a better school district. The three considerations for them are, what will it cost to move to another house instead of remodel? What would it cost to remodel? And are they looking forward to the remodeling process?

- They have seen houses that sell for $210,000 that have the extra space that they need and are in nicer neighborhoods. Houses similar to theirs in their neighborhood have sold for $140,000. So a larger house, if they moved would cost them the expense of a larger mortgage ($210,000 versus $140,000) plus the expense of moving.

- They answered in the Remodel-or-Move Calculator that if they remodel, they will hire a general contractor, do none of the work themselves, and select average-quality products when they can. In this instance, that would mean average-quality windows and doors found in new homes, standard and raised ceiling heights, and carpet, tile, and hardwood for the floors. Using

Table 6.3 Remodel-or-Move Calculator—Major second story remodel example

Question	Answer
What could you sell your home for today?	$140,000
If you move to a house today that has all the features you need, what would it cost?	$210,000
What state do you live in?	Oklahoma
How long have you lived in your home?	6 years
If you remodel your home, how much longer will you live there?	15 years
Who will do the basic design sketches and ideas?	An architect
Will you prepare the design drawings and apply for the permit yourself or hire someone?	My general contractor
Will you manage the project yourself or will you hire a full-service contractor?	Hire a general contractor
How much of the work will you do yourself (electrical, framing, painting)?	None
What level of quality do you want in the finish work? Economy, lower-cost bathroom, and kitchen fixtures, average-cost bathroom and kitchen fixtures, or custom and expensive bathroom and kitchen fixtures?	Average
Will bathrooms be remodeled? If so, how many?	Yes, 1
Will bathrooms be added? If so, how many?	Yes, 1
Will the kitchen be remodeled?	Yes
If you are remodeling the kitchen, will the size of the kitchen change?	No

Table 6.3, *continued*

Question	Answer
Will a second story be added?	Yes
How many rooms will be added, excluding bathrooms, living rooms, and family rooms?	1
Will a living room be added or enlarged?	No
Will a family room be added or enlarged?	Added
Remodel cost estimate, low range	**$115,000**
Remodel cost estimate, high range	**$185,000**
Estimated project payback	**$92,000**

Table 6.3, *continued*

Sell your house and move calculator

Question	Answer
Will you sell your home yourself or hire a full-service real estate agent?	Hire an agent
Cosmetically, how would you rate your home today? Is it ready to sell or do you need to do some minor cleanup, repair, painting, etc.?	Good condition. It needs a little paint and other things fixed before selling.
Number of bedrooms in your home	3
Number of other rooms furnished or used for storage (family room, dining room, den, living room, basement, attic, sun room, recreation room, garage, basement)	4
Number of people living in the home	4
Number of years you have lived in the home	10
Will you hire a full-service mover or pack yourself and rent a truck?	Pack myself and rent a truck
Move cost estimate	**$15,000**
Expense of higher mortgage payments for the new house	**$31,000**

Table 6.3, *continued*

"Gut feeling" analysis

Question	*Answer*
Are you satisfied with your local schools?	Somewhat satisfied
How important is the quality of the schools to you?	Very important
How do you feel about the distance from your home to your place of employment and other places you visit regularly (grocery store, etc.)?	Unsatisfied
What are your feelings about the remodeling process? Are you excited about it or dreading it?	Indifferent
How important is the neighborhood to you?	Somewhat important
Do you like your current neighborhood?	It is okay
Do you like the floor plan in your current home?	It is okay
Is the size and condition of your home's yard important to you?	Somewhat important
Do you like the size and condition of your home's yard?	It is okay
Is your goal to live in your dream house, or will any comfortable house meet your needs?	To have a few special features in my home

Table 6.3, *continued*

Remodel-or-Move Calculator Results

Question	Answer
Estimated move cost	$15,000
Expense of higher mortgage payments for the new home	$31,000
Total cost to move	**$46,000**
Remodel cost estimate, low range	$115,000
Remodel cost estimate, high range	$185,000
Project payback	$92,000
Remodel cost above payback	$23,000 to $93,000
Expense of higher mortgage payments for remodeled home	$32,000
Total cost to remodel	**$55,000 to $125,000**
Is it more expensive to move or remodel?	It is $9,000 to $79,000 more expensive to remodel

> The cost to remodel is more than to move. Kurt and Marie's gut feeling score indicates a preference to move. The Remodel-or-Move Calculator recommends moving.

THE REMODEL-OR-MOVE CALCULATOR

this information, the Remodel-or-Move Calculator estimates that their remodel project, which includes a second story with a new master suite and a remodeled bathroom and remodeled kitchen, would cost between $115,000 and $185,000 minus the likely payback of $92,000 plus the cost of financing the $92,000.

Using the information that Kurt and Marie provided, the Remodel-or-Move Calculator determined the cost to move, approximately $46,000, and the cost to remodel, between $55,000 and $125,000. Since the cost to move is less than the cost to remodel, and Kurt and Marie would be happier in a different neighborhood and with different schools, the Calculator recommends that they consider moving instead of remodeling.

As with all recommendations by the Remodel-or-Move Calculator, it is very important to get quotes and advice from both real estate and construction professionals to determine the actual costs before making your final decision.

With the help of the examples and the Calculator from the Remodel-or-Move Workbook or online at www.remodel ormove.com, you should have made your decision to re-model or to move. If you have decided to move, turn to chapter 7, "So You Have Decided to Move—What's Next?" If after completing the Calculator you have decided to remodel, turn to chapter 8, "So You Have Decided to Remodel—What's Next?"

If the Calculator provided a recommendation that doesn't agree with your gut feeling, consider changing a few of your answers in the Calculator to understand what it will take to have the recommendation from the Calculator agree with your emotions. Here is a list of steps you can try to make yourself more comfortable with the results.

If the Calculator recommendation is to remodel and you feel like you should move,

- Review your answers in the gut feeling section. If you really feel like you should move, make sure your answers reflect that. Are you truly happy with the yard, neighbors, location, and schools?

- You should also verify that your answers in the remodeling section are correct. Maybe you want more done to your house than you answered. Double-check your answers, change a few in this area, and see if the Calculator recommendation is the same.

If the Calculator recommendation is to move and you feel you should remodel:

- Review your answers in the gut feeling section. If you really feel you should remodel, make sure your answers reflect that. Do your answers indicate you are satisfied or very satisfied with the yard, neighbors, locations, and schools?

- You should also verify that your answers in the remodeling section are correct. Maybe you listed too many rooms to be remodeled, or you selected too high a quality for the materials. Double-check your answers, change a few in this area, and see if the Calculator recommendation is the same.

7

So You Have Decided to Move—What's Next?

You decided to sell your house and buy a new one instead of remodeling. Congratulations! More than six million homes are sold and purchased each year in the United States, so you will be in good company! This chapter will give you an outline of what you need to do to prepare your home and yourself for the successful sale of your current home and purchase of your new one.

First, you have to prepare yourself for the move financially and emotionally. Selling a home can take many months after you make the decision to move. You need to decide if you want to sell it yourself or hire a real estate agent, you need to prepare your home, you need to find your new home, you need to close on both sales, and finally, you need to move. Financially, you will have to pay anywhere from 3 to 10 percent of your current home's value to sell it and another 2 or 5 percent to move into your new home. This means if you are selling a $200,000 home and buying a $300,000 home, the total cost will be $12,000 to $45,000. This is a lot of money and time to invest, so take the time to learn what you can about the process to ensure you get your money's worth.

Homeowners always wonder when is the "right" time to sell a home. Most homes are sold during spring and summer. This is because many schools are closed during the summer and families with children can enroll their children in the new school in the fall. While buyers have more homes to choose from in the summer, for sellers this means there are more homes on the market that are competing for the same buyers. The converse occurs in the winter months, when fewer homes are sold and there are fewer buyers. Homes are typically on the market a shorter period during the summer, but the prices paid are not significantly higher or lower than in the winter months. Therefore, there isn't a "right" time to sell your home. You should sell your home when you are ready for the investment of time and money.

Another pressing question is when should you buy your new home? Should you find your new home first then rush to sell your current home? Or find a buyer for your current home first and then rush to find a new home? Which is the right order of events depends on your preferences and on luck. Both situations are stressful, but you can be successful in both cases by preparing carefully before you begin the selling and buying process. Your schedule will be dictated to a certain extent by the buyers of your old home and the sellers of your new home. You may be fortunate enough to receive an offer from a buyer who has no set schedule for moving. In this case, you can rent back your current home from the buyer for a few months while you look for your new home. This can be true of a seller as well. Some sellers don't have an urgency to move so they are willing to accept your offer, close on the house, and then rent the house back from you for a few months while you sell the house that you have. In either of these cases, you can find yourself with enough time to find your new house or to sell your current home without too much pressure.

To make shopping for your new house as easy as possible, it is important that you make decisions about the kind of house you want to buy and the location as early in the process as possible. It is just as important to decide on what you don't want in your new house and in the new neighborhood as it is to decide what you do want. Review both the reasons to move and the reasons to remodel in chapters 1 and 2, and list what you must have in your new home and what you want to avoid. These are difficult decisions that shouldn't be left to the last minute. With this list in hand, mark a street map indicating the areas that you will consider moving to and the ones you will not. With a list of what you want from your new home in hand and a map of the suitable locations, you are ready to find your home.

You can work with a real estate agent to find your home. Choose one that has a lot of experience in the neighborhood you have selected. This is important because a real estate agent who has spent years finding buyers for homes in a specific neighborhood will have a number of contacts, both homeowners and other real estate agents. With these contacts, they can help you search for your home, possibly finding a home that you can buy that is not yet on the market. You can also shop for houses yourself by driving through your targeted neighborhoods and directly contacting the sellers of homes that have for-sale signs.

The Internet has made this search process easier by offering comprehensive databases of all homes for sale. Some of these sites are listed in the appendix of this book. You can search for your new home every day by going to these Web sites and typing in the type of home you are looking for in a specific location. The Web site will furnish you with a list of homes, many with photos and a lot of detail. When one house looks good, you can jump in your car and drive by the house to take a closer look. If it still looks good in person, make a quick call to your real estate agent or the number on the for-sale sign, and you can be taking a tour in a few minutes.

Checklist for Preparing Your Home for Sale

Getting your house ready before you put it on the market can help you sell your home quickly and for a better price. For tips and strategies on how to make your home appealing, both inside and out, refer to "Getting Your Home Ready to Sell" in chapter 3 for details on how to proceed.

Deciding How to Sell Your Home

Once you and your house are ready, the first decision to make is, do you want to sell your home yourself or hire an agent? Most home sellers use real estate agents. They offer expertise in marketing your home, negotiating contracts, and preparing all the paperwork. For their services, they typically charge 6 percent of the sale price. More homeowners are choosing to sell their own homes. A real estate agent's services may be a good deal for a $200,000 home ($12,000 commission), but for a $600,000 home ($36,000 commission), the services become very expensive for some.

Reasons to Hire a Real Estate Agent

For many good reasons most homeowners use a real estate agent to sell their homes. The following list describes the most common reasons. A growing minority of homeowners choose not to use a real estate agent. Instead, they hire a number of professionals that each provide one or two of the services of an agent. Homeowners who sell their homes without agents do it to save money as well as to feel more control over the sales process.

Setting the Sale Price

An experienced, successful real estate agent will be able to help you select a sale price that will enable you to sell your house faster and potentially for more money. Qualified agents have the experience of selling homes in your area and understand what pricing strategies work best. Some areas require a medium sale price, which attracts buyers but offers little room for negotiation. In other areas, the pricing strategy may be to offer a low price and allow potential buyers to bid it up. A good real estate agent's experience in this area could be very valuable.

Staging

To sell your house fast and for the most money, the appearance of your home is very important. Having an experienced, critical eye look through your house and make suggestions for repairs, furniture layout, and decorating can add thousands of dollars to your home's sale price. The value of a critical eye cannot be underestimated. After living in your home for several years, surrounded by your cherished possessions, it's hard to be objective about which ones may take away from the appearance of your house for a typical buyer.

Marketing

A real estate agent has ready access and experience in the use of several very valuable marketing tools including the Multiple Listing Service (MLS), print advertisements in newspapers and magazines, Internet listings, and a network of other real estate agents. You have access to most of these marketing tools without a real estate agent, but a real estate agent has expertise on how and when to use these marketing tools as well as how to get exposure for your home faster.

Negotiating

When you sell your home yourself, you need to be able to act like an impartial negotiator not a proud homeowner, to maximize your sale price and profit. An experienced real estate agent can be a buffer between you and the turmoil of dealing with the buyer directly. Plus, a successful real estate agent can offer negotiating skills that most home-owners do not have. These skills could result in a higher price for your home.

Closing

Once you have accepted an offer, the work continues. There will be inspections to manage, financing to arrange, and paperwork to complete. Often, the inspections result in questions and new requests from the buyers. A good nego-tiator can manage the requests prompted by inspection findings to minimize the cost to the seller and ensure that the sale of your home closes.

Selecting a Real Estate Agent

Select an agent based on referrals, making sure that the agent has experience in your neighborhood so he or she will know the benefits and features of your neighborhood. An expert on an area of town thirty miles away won't know the best way to market and sell your home. Will agents from across town be as willing to meet prospective buyers at your home as they would if they did business a few miles away?

You should interview prospective agents to ensure they have experience in your neighborhood and homes in your price range, both selling and listing homes. This should be recent experience, within the last six to twelve months. Otherwise, they may not be up to date on changes, may be more focused on other parts of town, or may not be successful agents.

Selling Your Home Yourself

A real estate agent is not required to sell your home. More and more homeowners every year are choosing to sell their homes themselves. They choose to do it themselves to save a percentage of the value of their home from the standard 6 percent commission charged by real estate agents. On a $100,000 house, that savings will equal a few thousand dollars, but the savings on a $400,000 or $500,000 home can be five, ten, or even twenty thousand dollars. Some owners also appreciate the greater sense of control that they have when selling their own home. It is a major financial transaction, and some homeowners feel better having detailed knowledge and control over how the deal is conducted.

It is also getting easier to sell your home without a real estate agent because of the widespread use of the Internet. The Internet helps FSBO (for sale by owner) sellers have ready access to the information that they need to sell their home, including services that provide marketing tools, legal help, and general how-to information. The Internet has also enabled more and more buyers to search for homes on their own. They find houses offered by real estate agents and directly by owners equally appealing.

If you already have a buyer for your house, selling it yourself with some legal help from a real estate attorney is a practical way to save money. If you don't have a buyer, then consider carefully before attempting to sell the home yourself. The process has several pitfalls, including marketing your home, negotiating with potential buyers, and ensuring all the legal aspects are completed.

Determining the Price for Your Home

Whether you sell your home with a real estate agent or on your own, you will need to set the asking price. A good goal that most of us share is to sell our home for as much as possible. But how much can you get? The sale price of your home depends on several factors, including interest rates, the neighborhood, the overall economy, and the size and condition of your house. The simple way to evaluate all of these influences is to look at recent sales of comparable houses in a market analysis report. Many excellent sources on the Internet provide price estimates for your home. See the appendix for a list. If you live in a neighborhood with only one or two floor plans and several homes have sold in recent months, then determining the value of your home is straightforward. You need to look at the sale prices of the equivalent houses sold.

If your neighborhood is full of unique houses of varying sizes, then the job of determining the right sale price is more difficult. Successful real estate agents have a lot of experience setting selling prices, and some are skilled at it. However, don't assume that an agent has a secret for setting a price. It is an art, and no matter how much experience anyone has, mistakes can happen.

Once you determine what you think you can sell your home for, you have three options for setting the asking price. First, set it at the price you think it's worth. This is a safe, sane, and less stressful way to approach the pricing decision. Another option, which is sometimes popular in hot markets, is to price the home low in hopes of sparking a bidding war. The idea is that once buyers make an offer they have emotionally invested themselves in your home. Essentially, they feel like they have already bought it. When another buyer offers a higher price, the multiple bidders keep bidding on the price, resulting in a higher price than would have been paid if the asking price had been set higher to begin with. Many real estate agents like this approach because the seller tends to think that getting any price higher than the initial asking price, which was set low to start a bidding war, is a good deal, so the agent can sell the house faster.

Offering your home for less than the maximum price gives you a better chance of selling it quickly and painlessly. If it is a good deal, buyers will have fewer contingencies and may overlook small issues. Setting the price too high can damage your home's salability. If the price is too high, your house stays on the market longer. Buyers and their real estate agents may assume that something is wrong with it and not even consider making an offer.

The following is an example of a situation where a home was priced too high initially and then stayed on the market for months until the owners accepted less than it was worth because they needed to sell it.

> Terry and Melissa bought a home in Northern California in 1996. The market was slowly recovering from several years of falling prices. One house in their targeted neighborhood had been on the market for

three months, longer than average. The initial asking price was $322,000, too high for the condition of the home, which had an outdated kitchen and needed many cosmetic improvements. After dropping the price two times, the owners finally accepted $289,000 because offers had stopped coming in and they needed to sell the home. The owners would have sold the home faster and for a higher price if they had fixed the cosmetic issues by painting, cleaning, and decorating and had started at a more reasonable asking price of $305,000.

Accepting an Offer

If you have priced your home well, it is in good condition, and buyers are active in your area, then it is likely you will have multiple offers for your home. Congratulations! Multiple offers allow sellers to select the offer that works best for them instead of accepting all the terms requested by a single potential buyer. Offers have many points to compare. The obvious one is price. More is better, but consider the rest of the offer before making your choice. How are the buyers going to pay, and are they qualified for a loan in your home's price range? What other requirements do the buyers have? Do they want you to fix the water heater or the roof? Do they have other requirements? Consider all the offers carefully. Pick the one that best meets your needs and accept it. You're done. Now it is time to pack and move.

Packing and Moving

If you have decided to hire a professional mover, begin interviewing as soon as possible. Many Web sites have suggestions and recommendations on the best movers for your situation. See the appendix for suggestions. Regardless of whether you have hired a full-service mover or are doing it yourself, the following information will help make the process much easier.

There are two absolutes to moving: it is never too soon to start packing, and you have possessions that you should sell, donate to a charity, or throw away instead of move. You can start packing the seasonal items as soon as you can. These include holiday decorations, seasonal clothing, special occasion tableware, keepsakes—all of the things you won't need to use between now and the date you move into your new home. It helps to have a staging area where you can temporarily place the boxes that have been packed waiting for moving day. As you pack these items, ask yourself, "Do I really need this?" While it is often easier to pack things than to get rid of them, remember you will not only be packing the item, but you will also have to carry it (or hire someone to carry it), unpack it, and find a place for it in your new home. If in doubt, throw it out!

Purchase or locate new or used standard size boxes to use for packing your belongings. Having boxes of the same size makes stacking much easier. Get special boxes for clothing, books, china, and other heavy, fragile, or large-size items. You will also need packing material—bubble wrap or paper—as well as packing tape. A moving supply company can help with these items.

As you pack, make sure you mark each box with its contents and which room it should go in as well as a unique number, for example, "Pots, Kitchen, 23." Then record this information in a notebook along with more detail on exactly what is in the box, such as "large stock pots, frying pans." Note which items you want to unpack first. This will help with organizing when you are moving in. Plus, if you are missing any boxes after the move, you will know what was in them for insurance claims. Boxes with contents that are fragile or that must have a specific side facing up should also be labeled. Printed labels are easier to see and use instead of handwriting this information.

Unpacking and Getting Comfortable

The moving van just unloaded the last box and is pulling out of your driveway. You are now alone in your new house and eager to get comfortable and make the house your home. You look around for a place to sit and all you see are boxes, boxes, and more boxes. Now starts the unpacking stage. If you marked all the boxes with contents, room, and number, they should be in the right rooms or at least left in an area nearby to be moved later. Using the list you created when packing, start with the high-priority rooms first, typically the kitchen and a bathroom. As you empty boxes, flatten them and consider recycling them. Many moving companies and packing material companies will take back the boxes you used. They will sell them as used to the next family that needs to move.

After you have settled into your new home and begin to enjoy the added room, larger yard, better schools, or other features that motivated you to move, you can have the confidence that you decided to move instead of remodel because it was the right decision both financially and emotionally for you.

8

So You Have Decided to Remodel—What's Next?

Y ou decided to remodel your home instead of selling and moving. Congratulations! This chapter will give you an outline on how to prepare yourself for a successful remodel project both financially and emotionally. Remodeling a home can take many months after you start the process. You need to decide what you want to remodel, what you can afford, how to pay for it, and whether to manage the project yourself or hire a general contractor. The list of decisions goes on and on. Making these decisions is much easier if you are organized, so set up a filing system immediately, following the outline below. Good luck. Try to enjoy the remodeling project as much as you will your home once the work is done.

Getting Organized

An unbelievable number of decisions need to be made during a remodeling process. For most people, this is a daunting task. To make the job easier and to ensure your percentage of good decisions is as high as possible, you need a strategy to organize the information. You can use a three-ring binder or an expandable file folder, whichever you prefer. Make sure it is durable and easy to carry because you will want to take it with you when you are shopping for materials and whenever you are out, in case you need to quickly contact someone that is working on your home. Start with the following sections and customize the folder as you go. While most of the organization system is optional, keeping copies of all contracts, change orders, invoices, receipts, and permits is a must. These will be invaluable if you have disputes with your contractors and, when tax time comes around, to calculate your investment in the remodel. Here are the sections you can start with:

Ideas

These are the photos, magazine articles, and notes on the way you want the finished project to look. Have a section for each room: bathroom, den, master bedroom, and so on.

Quotes

Save the brochures and quotations from candidate contractors and suppliers. You may also want to create summary sheets of the quotes you have received for the same items so you can quickly compare them.

Contact Information

Each contractor will have a mailing address and several phone numbers (cell, home, office, fax, pager) so keep them handy and in one place. Since most of this information will be on business cards, invest in a business card holder that fits your binder or folder.

Purchase Orders, Contracts, and Change Orders

Once you have signed a contract or a purchase order, you need to keep a copy handy for your reference. If you have questions about price, delivery, or warranty, you can quickly review the document again. Change orders are for changes that you and the contractor agree to make during the project. Changes happen in all projects, so be prepared for them and make sure that all the details of the changes are in writing and signed by both you and the contractor.

Invoices and Receipts

Keep records of all your expenses so you can total them at the end of the project to see if you hit your budget or not. You can also use this information later to calculate your cost versus any gains you make on the price of your home to determine your capital gains tax.

Permits

Each permit-issuing authority has rules on how to make the permit available to inspectors. Follow those rules, and after the project is done, store your permit in the binder or file.

Listing What You Want

Now that you have a system to organize your remodel project, start by creating a list of what you want—a bigger bathroom, new kitchen cabinets, a fourth bedroom, and so on. Once you have the list, group the items by room type. Sort them as well by whether they involve remodeling an existing space or creating a new space. A sample of a remodeling goal list is below.

Table 8.1 Remodeling goals

Remodel existing space	Added space
New kitchen cabinets, floor, appliances, and countertop	Master suite
Built-in cabinets and fireplace in living room	Living room expansion
New closet doors	Entry hall coat closet
New tub, tile, and vanity in bathroom	

Once you have this list, try to get a few specific details. Photos from design magazines, a sketch of the layout of the room additions, and decisions on the type of materials, fixtures, and cabinets you want will help you estimate the cost and time required to complete the project.

Creating a Seamless Addition

You have the list of what you want. Now carefully consider adding to it. This is important for four reasons. First, you will be able to make a better decision about what is most important if you have a complete list of every possible project that you would like to accomplish. Second, ensure that your changes are seamless with the existing house. You may want to add little details throughout the house to ensure a consistent look and feel. Add to your list the items that are needed in each room and on the exterior to make your home look like it was built with all the additions and renovations already in place. Your goal is to create a seamless addition where a casual visitor couldn't tell what was added or what was original. Third, there may be changes that you could make for little or no cost while other work is being done; unless you have them on your list when you start, you may miss an opportunity. Finally, you want your list complete with all possible projects so that after you start construction, you will be less likely to have any additions or changes. Midproject additions and changes are the biggest reason that some projects run over budget and are completed late.

Some areas need to be reviewed carefully during your planning phase to ensure your addition and renovations fit in with the existing construction. The doors, windows, and hardware in the addition and the existing construction should match or complement each other. It may be a good idea to replace the doors and windows in the existing construction if you are changing the style of the home. The roof material and pitch as well as the exterior finish of the home should match or complement the existing finishes. Although it is okay to mix floor types (carpet, tile, and hardwood), when you have two of the same type, ensure that they match or complement each other. The wall texture, finish, and ceiling heights should match as well. Having one room with a smooth-textured 9-foot ceiling, which is common in newer homes, when the rest of the house has 8-foot ceiling with a rough finish will disrupt the overall feeling of the home, unless the contrast is part of the look you are trying to achieve.

The Design

Many "amateur" home designers come up with great ideas, and since you are living in your home day after day, you can have better ideas than the "pros." Get a pencil, make 100 copies of your current floor plan, and sketch out all the alternatives. It doesn't matter how good or bad they are; experiment with alternatives. At this point it's free, so enjoy!

You can get a drawing of your current floor plan by making a sketch yourself. You can do this freehand, or you can purchase a computer program to make the resulting drawing neater. Several software programs are recommended in the appendix at the back of this book. You can also hire an architect or a home designer to create a floor plan for you.

As you get closer to making your final decision on your plans, reality needs to come into play. You are almost always better off sticking with a traditional size and shape in a home, both for resale value and cost for your remodel. This means room sizes should be close to standard, not too large or too small. Doors and windows should be sized so that they are easily replaced, and floor plans should be traditional as well. See the appendix for a list of resources for sample floor plans.

Having a kitchen on the second floor may strike you as a clever and unique solution, but if it means moving the gas, electrical wiring, and plumbing up to the second floor, this change will add significantly to the project costs. You should also consider resale value. You may plan to stay in your home forever, but the unexpected does happen, and if you installed bright orange tile on all the floors, most perspective buyers will deduct the cost to replace it from their offer even though you installed it just two months before.

Deciding What You Want to Do and What You Want Done

The following is a list of the different phases of the project. You can decide to do some of them yourself, which will save you money, or hire someone else to do them for you. Refer back to the section about the major tasks in a remodel in chapter 4 for a ballpark estimate of the potential savings. Another good way to determine the potential savings is to have contractors itemize their quotes. If you request this, the contractor will give you a quote with specific prices for foundation, framing, electrical work, and so on so you can see your potential savings if you take on these tasks for yourself. Take time, research the work that you may be interested in doing, and make the decision early. Changing your mind about wanting to do this work yourself after the project is under way is not a good idea. Frequently, contractors will give you a competitive price on the whole project, but when you make changes, they will charge a higher rate because customers have few options once the project is started.

Project Management

Hiring and coordinating the different trades is a reasonable task for most homeowners to undertake. If you have organized different professionals before, such as planning a big wedding or managing employees at work, you can do this. Savings can be 5 to 50 percent of the total remodel cost. This is a good task to research and, if you feel confident, undertake.

Design

If your project is small, just a bathroom remodel or something similar, then you can design the project yourself. (You may even enjoy it.) If it is a bigger project, you can still do most of the work and save 5 percent of the total project cost. You will probably need to have an engineer review your design and add the necessary technical details to get the permit. The engineer may also clean up your drawings so they conform to the permit and construction standards.

Permit

Getting a permit from the city or county is something you can do, but there is little cost savings for you, and the process isn't particularly enjoyable. It is similar to most interactions with government bureaucracies. If you aren't hiring a general contractor, then you should get the permit since you will be the only one familiar with all of the aspects of the job.

Demolition and Preparation

You can do the demolition and site preparation, and many people enjoy the fast progress of this work. However, there is little savings for you since it is work that the least expensive laborers do. In addition, demolition and site preparation represent only a small percentage of the entire project cost.

Foundation Work

Pouring concrete walls or slabs is not something most do-it-yourselfers should undertake. Also, with a little shopping around you probably can find very competitive bids for the foundation work.

Framing

Framing the floor, wall, ceiling, and roof structure is a fast process that requires strength and physical effort as well as an average amount of skill and training. Building a straight wall (square, plumb, and level) is straightforward work. Building a complex valley on a roof two stories up can be a challenge. If you have the time, tools, and interest, you can consider doing this work.

Electrical

This is one of my favorite do-it-yourself projects for several reasons. First, contractors typically charge the highest rates for this work, so you can save a lot doing it yourself. Second, pulling the wires through the walls and ceiling and installing the boxes and fixtures is not too difficult. Finally, the work goes fast and is rewarding. However, I would not recommend installation of a breaker box or connecting your home to the electrical service. Hire an electrician who can also look over your wiring while he or she is at the job site.

Plumbing

Plumbing can be a do-it-yourself project, especially if it involves all new installations instead of modifications to existing supply and drain lines. Consider carefully before taking plumbing on as a do-it-yourself project. Generally, I would not recommend it unless you have experience or a lot of time and interest.

Heating and Cooling

Installing the heating and cooling vents and ductwork is a good do-it-yourself project. The ductwork currently available is flexible and pre-insulated, making it hard to go wrong. Hire a professional to install the heating and cooling unit and to hook up the electrical and gas lines. Then you can do the rest.

Interior and Exterior Surfaces

Installing drywall inside or siding and shingles outside is frequently heavy work that requires some skill, so unless you have done this type of work before, I would recommend hiring a professional who has the tools and skills. The hourly rate charged by this group of tradespeople is reasonable, and they will work a lot faster than a beginner will.

Finish Flooring

Prices charged by professionals who install hardwood and carpet are often reasonable so I typically recommend that homeowners shop around and hire a professional. I have found that a homeowner can do a good job installing tile with practice and the right tools. If you are interested, have the time, and want to learn a new skill, then save yourself some money and install the tile yourself.

Doors and Windows

Windows can be installed by the same professionals who do the framing. Doors can be installed by the finish carpenters. Both of these tasks can be done by a homeowner successfully, so consider them if you are interested in a small cost savings.

Cabinets, Fixtures, Appliances

You should consider installing fixtures and appliances if you are handy around the house. The work goes fast. You need to focus on the details, but if you are good with instructions, it is likely you can do a good job installing cabinets and appliances. Savings can be up to 30 percent of the project cost.

Finish Work

Another good candidate for a do-it-yourselfer is performing the finish work. Installing baseboards, doors, crown molding, and other finish trim requires a moderate amount of skill, a few special tools, and a good attention to detail. If you meet these requirements, then doing the finish work can be a good project for you to do. Savings can be up to 5 percent of the project cost.

How Much Can You Afford?

Once you have your list of possible projects, you will need an estimate of the project cost. Then you will need to look at your budget. How much do you have to spend on the remodel? How much do you want to spend?

How much will the project cost? Using the Remodel-or-Move Calculator in this book will give you an idea. But for a more accurate figure, you'll need to prepare a detailed estimate if you are going to do most of the work yourself or ask for several bids from contractors. You should also plan to spend at least 20 percent more than the contractor's bid on unexpected expenses. It is better to plan to spend a lot and spend less than the opposite. Also, if you are doing most of the work yourself, be very careful with your estimates. Some people are very good at estimating; others aren't, so you need to understand which type you are. I think I am a very good estimator, but when I estimate my own projects, I am optimistic on the time and money involved. My wife knows better. On small projects at our home, she doubles the costs that I tell her and triples the time it will take. For bigger projects, she is a little more trusting of my estimates, adding 50 percent to the cost and 50 percent to the schedule. Whatever system you use, make sure that you are overestimating the time and money to save yourself many headaches.

Table 8.2 Example of a simplified remodel budget

Project: 123 Main Street Remodel

Contractor price quote	$65,000
Unexpected expense allowance (20 percent of contractor's bid)	$13,000
Total cost budget for remodel	**$78,000**
Contractor schedule estimate	8 weeks
Allowances for construction delays, preparation prior to construction, moving back in, and decorating after construction is done	6 weeks
Total duration of the remodel project	**14 weeks**

Once you have an accurate estimate of the project cost, you should check your finances and make sure you can afford your remodeling projects. Remodeling projects are infamous for going over budget. It doesn't happen for some mysterious reason. It happens because important details have not been planned thoroughly, the budget was too low for the project, and decisions during the project are made that drive up the project cost. Therefore, before you decide on your remodeling plans, understand how much you want to (or can) spend.

How to Pay for Your Remodel

Financing a remodel is the most common way to pay for larger projects. Typically, the lowest-cost solution is to take a home equity loan. This makes a lot of sense since the dollars are invested back into the house and the interest that you pay on the loan may be tax deductible. Construction loans are also available. They are best suited for very large projects costing more than the equity you have available, such as adding a second story or totally tearing down and rebuilding your house. Construction loans are just for the life of the project. They are typically paid off by a standard mortgage on the home after the project is done, which is good since construction loans cost more than standard home mortgages. Check with your bank or credit union for options available to you, as well as for competitive interest rates.

Credit cards can also be used to finance a remodel project, but they need to be managed carefully since they have a higher interest rate, typically more than home mortgages or construction loans. Loans against a 401(k) or similar retirement account are a relatively new alternative for most of us. They may or may not be available to you depending on the 401(k) program you participate in. A disadvantage is that if you lose your job, most 401(k) plans require that these loans be paid off immediately. This can be a little scary if you are in the middle of a remodeling project. You could not only lose your job but also have to come up with a large amount of cash to pay off a loan!

Finally, loans are available from some contractors. These are typically not the most competitive loans. In addition, a loan through a specific contractor could lock you into using that contractor. If you aren't happy with the

Table 8.3 Comparison of ways to pay for a remodel

Type of loan	Pros	Cons
Loan against retirement account [e.g., 401(k)]	You pay yourself the interest on a loan.	You lose the interest you could be making if the money was invested. If you lose your job, most loans require you to pay the loan back immediately.
Home equity loan	Usually tax deductible. Lump sum is paid to you at the start so you have flexibility of what you do with the money.	A second loan to manage. Shorter term than a standard mortgage. Requires that you have sufficient equity in your home. You have to pay interest on the entire loan amount even though you may not need the money to pay for remodeling right away.
Home equity line of credit	You borrow only the money you need at the time so finance charges are lower at the beginning.	A second loan to manage. Shorter term than a standard mortgage. Requires that you have sufficient equity in your home.
Construction loan	Good for larger remodel projects and if you don't have enough home equity to take a loan to cover construction costs.	Higher interest rate than home equity loans. Not tax deductible. Usually short-term until construction is complete then replaced with a new first mortgage.

Table 8.3, *continued*

Type of loan	Pros	Cons
Loan from the contractor	Available to most homeowners.	High interest rates. Not the the best terms. Can lock you into working with a specific contractor. Not recommended.
Refinance and cash out	You have only a single loan for your home. Usually tax-deductible interest. A single larger loan will usually have the lowest interest rate.	Requires that you have sufficient equity in your home. You have to pay interest on the entire loan amount even though you may not need the money to pay for remodeling right away.
Credit cards	Most homeowners have this as an alternative.	High interest rate. Not tax deductible.
Your savings	The least expensive way to pay for your remodel.	Make sure you don't use all of your savings. Always have some available for emergencies.

contractor's performance, your options are very limited. You are usually better off getting a loan from another source.

How to Pick a Contractor

You have the choice of hiring a general contractor to do the complete job or hiring subcontractors with the different skills directly and performing the project management yourself. In either case, there are several things to consider when hiring the tradespeople to perform the work for you.

First, make sure the contractor is fully licensed with local and state governments. You should check with your local jurisdiction to find out if a license is required. If it is, ensure the license is valid and up to date—do not rely on the contractor's word. You should contact your state, county, and city governments for information on the contractor's license. Licensing requirements for all fifty states are available at www.contractors-license.org.

Make sure the contractor is fully insured and carries workers' compensation, liability, and property damage insurance—ask for copies of the documentation. In addition, check with your insurance agent so you are aware of what your insurance covers and what it does not.

Once the above details have been established, explain your project, review the drawings, sketches, and plans, and ask the contractor's opinion on the project. This is the best way to gauge the contractor's interest and capability. You want to pick a contractor that has experience with projects very similar to yours. You don't want to pay for an expert and get a novice.

Ask what work the contractor will do him- or herself, what the contractor's employees will do, and what jobs will be contracted out to others. It is best to hire contractors who will do most of the work themselves and only contract out some of the specialties, such as electrical work, tile, or plumbing. If the person you interview is the one who is going to do 80 percent of the work, you will have less chance of miscommunication than if you work with a general contractor that hires others to do all the work.

Ask for local, recent references as well as a history of the contractor's experience and business. An individual with ties to your community is usually best. When you check references, ask about the contractor's work habits, the timeliness of the work, and the quality of the results. Would they hire the contractor again?

How to Find a Contractor or Subcontractor

Finding full-service contractors is easy. They have big advertisements in the yellow pages, in the newspaper, and on the Internet. But just because contractors can afford to spend money on big advertisements doesn't mean that they are good, so make sure you get past the brochure and the sales pitch and go through the steps outlined above. You don't necessarily want a large contractor or a small contractor; you want the best contractor for you and your project.

Finding a smaller general or subcontractor takes a little more effort. They sometimes advertise in local newspapers— the community ones that are distributed for free. Try your local monthly or weekly city newspaper. The contractors who advertise here have an interest in your city and are focused on the area. The last thing they want is an unhappy customer in a town of 50,000 people. Also, ask around. Although you should take recommendations from other homeowners with a grain of salt, recommendations from other subcontractors often are worthwhile. They usually only give you names of contractors who they have worked with and like doing business with. These are the ones you want to work with too. So, the next time you need some painting done or a little concrete work or gardening, ask the people you hire for the small jobs whom they would suggest for bigger jobs.

Contracts

Most people, including contractors, are honest and hardworking. They want happy customers. Your job in the contract process is to avoid the few contractors who are dishonest. Therefore, structure your contract so it is easy for the contractor you hire to do a good job and make you happy. The bad experiences you hear about do happen. In many cases, it is the contractor's fault, but in many other cases, it is the homeowner who has unrealistic expectations. Make sure you have reasonable expectations. You are spending your savings and building your dream home, but it is just another job to the people doing the work. They have good days and bad days. Also, remember, that it is a house. It is not a work of art or a new car, which is precision fitted and near perfect. You want quality work, but probably don't want to pay for (nor will you get) perfection. Put everything in perspective as you go through this process and remember it is just a house. Small imperfections may be disasters to you but will likely be unnoticeable to anyone else.

Your contractor will usually have a standard contract for you to sign. Read it and understand it. Consider having an attorney review it. Most contracts are fine. You should have as many of your instructions and specifications in writing as possible. The specifications should include the tasks to be performed, material types, who supplies them, and when the project will start and stop. Ideally, segment the different sections of the contract so they conclude with an inspection by your local government and specify that the work has to pass inspection. This will give you a higher level of confidence that the work was performed adequately. Include notes on who is responsible for cleanup and dump fees.

Make sure all agreements and all changes are in writing and initialed by you and the contractor. Do this for even the trivial items. You probably won't ever need to take legal action, but all of us remember things a lot better if we read them and sign a document than if we are just told. Therefore, to ensure your contractor has a good memory, get everything in writing.

When you get a bid, make sure it is complete and provides enough detail. If you want more detail in the bid, feel free to write an addendum with the details you want and have the contractor review and sign it. Your primary goal is to hire someone who will do a good job at a reasonable price, not necessarily one that has the most complete contract. In addition, many great contractors aren't great business managers, so spend the time to add details to a contract from a contractor you like. If your requirements are reasonable, a good contractor will agree.

You need to trust the contractor; make sure that the person who quotes the project and explains the contract to you is the same person who will do the work. You don't want to meet and sign a contract with one person and then find out someone else will be doing the work. You should never be rushed into signing a contract. Take your time, review it carefully, and sleep on it a few nights.

Consider the payment terms carefully. Any advance payments should be for materials that are delivered to the work site only. In many states, this is limited to 10 percent of the total project expense. Additional payments after each phase of the work is done are appropriate and needed for most small contractors, who have weekly payrolls and materials to pay for. The final payment of at least 10 percent should not be made until all issues have been resolved. It will take several days to ensure everything was done to your satisfaction and to get the final inspection by the city. Don't be rushed by the contractor. Take your time and make sure the work was done right and that all final releases of the lien and a copy of the final invoice showing that the contract has been paid in full are signed. After you pay the contractor and he or she has started a new project, it will be difficult to get the small problems repaired.

It may also be wise to add a note in the contract that if the project is not finished within a specific time, deductions will be made from the total project cost. This ensures that the contractor will take the schedule commitments seriously. If you have a deadline for a holiday or for a family gathering, start the project well ahead of time so that the scheduled finish date is at least six weeks before your target date. Then agree with the contractor that if the job is not finished by that date, $100 or $200 per day will be deducted from the project total. If you don't have a specific completion date requirement, take the schedule estimated by your contractor and calculate an end date. Then add two weeks to this date and specify that the project must be completed, or deductions will begin. In addition, stress that complete means *complete,* not almost complete. This includes the minor touch-ups, light switch adjustments, and cleanup; it all needs to be done to your satisfaction and generally accepted workmanship standards. You need to be fair to your contractor and have reasonable expectations. Agree in writing on the completion date and then don't change the project requirements. You also need to be fair to yourself. If the job isn't complete, make a list of what needs to be done and give it to the contractor on the day that he or she agreed to finish and begin deducting the agreed-upon amount from the total still owed. The project will be finished, with a reasonable level of workmanship, before you know it.

Contract Checklist

Every contract should include the following:

- Contractor information including name, address, telephone numbers, and license number.

- A list of what the contractor will and will not do. What contractors won't do is as important as what they will, so make sure these items are included. Examples are the tasks that you have chosen to do, such as painting or demolition.

- All materials, size, colors, specifications. For example, windows: vinyl, wood, or aluminum, specified by manufacturer and model number.

- A dated copy of all drawings and diagrams. If changes need to be made during the project, they should be made to these documents and be initialed and dated by both you and the contractor.

- Start and finish dates.

- The times work will start and finish and the days of the week that workers will be at your home.

- How change orders will be handled.

- A warranty for one year.

- A binding arbitration clause.

- A statement of how the contract can be canceled.

- A statement that the contractor will provide affidavits of final release, final payment, or final lien waivers from all subcontractors and suppliers.

Liens

A construction lien allows any contractor, subcontractor, or supplier to place a legal claim on your home as security for payment of a debt. The debt in this case is the payment for services or materials provided to you or your general contractor for your remodeling project. A lien against your property will inhibit you from selling or obtaining financing until that lien is paid. The most common liens occur when the general contractor fails to pay his or her suppliers. This can happen even though you have paid the general contractor in full. The general rule is never to make the final payment to the general contractor for the remodeling project without receiving a release of liens from the contractor, subcontractor, supplier, and whoever else is involved. In some states, it is required that subcontractors and/or suppliers notify you that they will be performing work and/or providing supplies. If your state does not require this notification, request it in the contract. Save these notices with your invoices and your contracts for reference if needed.

This chapter has provided you with an overview of the remodeling process. If you are comfortable with the process, then you are ready to start with your remodel. If you are still unsure about aspects of the remodeling process, many great books and Web sites can provide details on every aspect of the process, from deciding what you want to remodel to putting the finishing touches on the work at the end of the project. Some of these resources are listed in the appendix of this book. You can also meet with a contractor or an architect who can discuss the project with you and answer any questions that you may have. Many of these professionals are excellent advisors on the remodeling process and are often glad to help.

If you are ready to start your remodel process but are not sure if you want to do some of the work yourself, you can get some hands-on experience in several ways. The first is to find a project around your home that is small and not involved with the larger remodel and try it yourself. You might consider putting a floor in the attic to allow easier storage, building a shed in the backyard, or constructing shelving in the garage. If you ask around, you will likely find friends or neighbors who have small construction projects they need help with. Often community education programs offer courses on remodeling, and Home Depot, Lowe's, and other building material stores offer classes on various construction skills, such as tile installation and basic wiring.

You can also volunteer at a Habitat for Humanity project or other charitable organization that builds and remodels homes for people. Volunteering to help on one of these projects will give you a firsthand look at the work. You will meet a lot of people interested in construction, some of whom will have recommendations for subcontractors that you can use on your remodel project.

9

Frequently Asked Remodel-or-Move Questions

This chapter answers ten common questions that homeowners have about the remodel-or-move decision. These answers can also be found in other parts of this book, but to help you make the decision, the specific responses to the questions have been gathered together here.

How Much Will It Cost to Remodel?

There are several ways to estimate the cost of a remodel project.

Cost per Square Foot

The most common way to do a quick estimate is to simply assign a cost per square foot to the area that will be remodeled. To calculate your own estimate, determine the square footage of the area that will be built new or that will be substantially changed and multiply that by a cost of $100 to $700 per square foot.

For example, if you want to add a 150-square-foot bedroom at the end of a hall by building an addition into the backyard and substantially remodel an existing 150-square-foot bedroom and a 50-square-foot bathroom, the total area would be 350 square feet (150 + 150 + 50). Then take 350 and multiply it by the range of possible costs per square foot, $100 to $700, and you get an estimate for this remodel of between $35,000 and $245,000.

You can quickly see that this estimate has little value since the range of potential costs is so great. While the $700-per-square-foot price is unusual and it is unlikely you could spend that much, the $100-a-square-foot price is likely unrealistically low unless you make great efforts to minimize the cost. The Remodel-or-Move Calculator gives a more accurate estimate by including information about the rooms that will be remodeled, the level of finish that you want, how you will manage the project, and where your home is located.

Estimate from a Contractor or an Architect

With a general idea of what work you want done as noted above, you should be able to get an estimate of the cost from a contractor or an architect. These professionals will likely use cost per square foot to calculate an estimate, but they will not use as large a range as $100 to $700. Instead, they will use an average that is based on their experience with similar projects in the area where you live and your requirements as to the types and quality of materials to be used. Their estimate will be much more accurate than an estimate strictly based on the square footage. The Remodel-or-Move Calculator uses the same types of information as an architect and contractor, and because it is fully automated, it allows you to try a number of different combinations of materials and room changes for free.

What Should I Do If the Contractor's Quote Is More Than I Can Afford?

It is common for homeowners to have "sticker shock" when they get the first quote for their remodeling project. After months of planning and anticipation, you have finally decided what you want done. You meet with the contractor to discuss the details, and a week later you get the quote. You are surprised to find the price is twice as much as you were expecting and had budgeted for. Now what do you do?

- Get more than one quote. All contractors have their own way of quoting and their own cost structures. Their quotes will vary. Sometimes they vary by a few hundred dollars, other times by hundreds of thousands of dollars. So get a few more quotes and see if there is a trend in the estimates.

- Confirm that the contractor understands the scope of the project. Unless you gave him or her detailed architectural drawings, there are many opportunities for your description of the remodel project to be misunderstood by the contractor.

- If the other quotes that you get are in the same range and you still feel that it is more than you want to or can spend, you can try the following:

  - Reconsider the level of quality that you requested. Also ask the contractor for ways to reduce the cost. Standard-size doors and windows and less-expensive materials for the floors and walls in a bathroom can make a tremendous difference in the total price.

  - Consider reducing the scope of the project. Maybe you don't need a larger breakfast area or a second walk-in closet.

- Consider managing the project yourself. If you have the skills and interest, you can save a lot of money by managing the project yourself. Look back at chapters 4 and 8 and review the descriptions of managing a project.

Should I Stay in the House During the Remodel or Move Out?

Start by asking the contractor what he or she recommends. There may be stages of the project where you need to be out of the house. If the contractor says that you can stay if you want to, consider whether you will have use of the critical areas of your home during the remodel project: bedrooms, a bathroom, and a kitchen. Will they be heated and cooled during construction? Will there be intermittent power and water interruptions? Also consider how much inconvenience you can tolerate. Remodel projects rarely go as planned, and even though you were assured that your bedroom would stay intact, a crack in the drywall or a hammer through a wall can happen anytime that workers are on a job.

How Long Will the Remodel Take?

Your contractor will give you an estimate of the time required to complete the work. But remember that the unexpected does occur during remodeling projects and there may be a project that the contractor has to finish before he or she can work on yours. So assume that the project will start late and finish late. It could just be a day late or it could be weeks. If you assume the worst, then you will be pleasantly surprised when it happens on schedule. Also, many homeowners are surprised that work isn't being done on their house every day. There are many reasons for stopping and starting projects, such as waiting for inspections, waiting for materials, working on other jobs, and workers calling in sick. So be patient. If you picked a good contractor, the project will turn out great and may even be done on the scheduled completion day.

Can You Trust Contractors and Their Employees?

Contractors and their employees are regular people, no better and no worse. Most want to do a good job and will treat your property with respect. To minimize the chance for problems, set up an environment where opportunities for accidents are limited. Since the contractors will be going in and out of your yard and your home with materials and equipment, make sure that the pathways are clear and uncluttered. If you don't want things damaged, put them away. Close off areas that the contractor and their employees don't need to go into. Taping plastic over the entrance to a room or hallway and locking interior doors to areas that they don't need access to is a very effective way to minimize the mess in other parts of your home. If you have valuable or breakable items, consider packing them away or take them to a friend's or relative's house until the project is over. As stated before, accidents do happen, and if you have a valuable vase or picture anywhere in your house, then it is best to store it away until the remodeling is done.

Where Can I Find a Contractor?

The most common question for homeowners who are considering a remodel is probably "Where can I find a contractor?" Review the methods suggested in chapter 8. Here are some more tips:

- Ask for referrals from friends, neighbors, and co-workers.

- Ask other people who have done work for you in the past, such as your plumber or gardener.

- Search your local and free newspapers for contractors' ads.

- Try online referral services such as www.servicemagic .com and www.improvenet.com.

- Use the telephone book.

- Ask at lumberyards and hardware stores, especially lumberyards that cater to contractors.

Isn't Remodeling Risky?

Remodeling is no more or less risky than other major decisions that will cost you lots of money. If you plan well, pick a good contractor, and have reasonable expectations, the project will be successful with only a few, very minor problems.

How Long Will It Take to Sell My House?

How long your house will be on the market depends on the number of buyers for houses in your neighborhood and the price you are asking at the time you are selling your house. Talk with your real estate agent to get an estimate of the time it will be on the market. Also watch other houses in your neighborhood that are priced at the same level and are being sold at the same time of the year. If you priced your home competitively and it is in good condition, then it shouldn't take longer than is typical for your neighborhood.

How Do I Pick a Real Estate Agent?

Make sure that agents have recent, successful experience selling houses in your neighborhood, check their references, and then decide who you think will represent you the best and do a good job. It is always a plus if you like your real estate agent, but it is less important than how well he or she will protect your interests by selling your home for the highest price and ensuring that the selling process goes smoothly. Focus more on the agent's track record of success rather than on whether he or she is a pleasant person.

Which Offer Should I Accept for My House?

The decision to accept an offer is difficult because of the short time you have to make it and the different advice you will get. You need to consider the price the buyers are offering, the number and type of contingencies, the ability of the buyers to obtain financing, and the chances of getting a better offer. Your real estate agent will play an important part in reviewing these offers so take advantage of his or her expertise. Also keep in mind that real estate agents get paid when you sell the house. They have a big incentive to sell your house as fast as possible. Some agents will put a higher priority on selling your house quickly than on making sure that you are getting the best price. Most will not knowingly recommend that you accept an offer that is likely to cause harm, but they may not recommend holding out for a higher offer if you have an acceptable one in hand.

10
Conclusion

R emodeling and selling your home are significant events in your life. They are some of the most expensive tasks you will undertake, costing tens to hundreds of thousands of dollars. Because of the importance of these decisions, it is critical to be well informed. Like other important decisions, no one can make a good decision for you. You need to gain enough knowledge to be a good consumer. *Remodel or Move? Make the Right Decision* is a great start for gaining this knowledge, but it cannot provide all the answers, especially to the questions that are unique to your situation. So treat the reading of this book as a good beginning to making this very important decision. The next step is to talk to local professionals about your options. As you talk to real estate and remodeling professionals, keep in mind that they have vested interests in your decision. Their input will help you confirm what you already know or bring up valid considerations that you might have overlooked. In either case, this additional information will help you get closer to your final decision.

Good luck to you on your decision. Remember that it is only a house; you make it a home. Whether it's big or small, newly remodeled or run-down, enjoy its good points and ignore its faults.

And finally, this book was written for you, the home-owner that needs help making this decision, so please send me a note and let me know how you did. Send me your thoughts, ideas, comments, corrections, and recommenda-tions for additions to this book. I look forward to hearing from you at dan@remodelormove.com or Dan Fritschen c/o ABCD Publishing LLC, 1030 E. El Camino Real, #150, Sunnyvale, CA 94087.

Appendix

THE FOLLOWING SECTION LISTS examples of remodeling drawings, sample remodeling contracts, and a list of resources and references to help you in your remodel-or-move decision.

Sample Remodeling Contract

The following is an example of a remodeling contract. If you plan on remodeling, you should read this sample thoroughly and review it with your attorney to understand which terms and conditions are acceptable to you and which are not. If you do this before getting quotations from contractors, you can be sure to mention the terms and conditions that are acceptable to you so that the contractors can take that into consideration when preparing their quotations. If you would like to use this form for your remodeling project or need other legal forms, a good resource is www.abcaforms.com.

HOME IMPROVEMENT AGREEMENT SAMPLE

This form complies with professional standards in effect January 1-December 31

THIS AGREEMENT BETWEEN:

_____	**SAMPLE**
(Contractor's Name)	(Owner's Name)
_____	**SAMPLE**
(Contractor's License Number)	(Owner's Home Address)
_____	**SAMPLE**
(Contractor's Address)	(City, State & Zip)
_____	**SAMPLE**
(City, State & Zip)	(Owner's Business Address)
_____	**SAMPLE**
(Telephone-FAX)	(City, State & Zip)

CONSTRUCTION LENDER: Name and address of construction fund holder is:

SAMPLE

(Name And Branch Address Of Bank, Savings And Loan Assn., Escrow, Agent, Joint Control Or Other)

DESCRIPTION OF PROJECT (including materials and equipment to be used or installed): Contractor will furnish all labor, materials and equipment to construct in a good workmanlike manner (Describe Labor, Materials, And Equipment To Be Furnished):   **SAMPLE**

© abcaforms.com
May Not Be Duplicated

WORK TO BE PERFORMED AT (Legal Description And Street Address If Known):   **SAMPLE**

TIME FOR COMPLETION: The work to be performed by Contractor pursuant to this Agreement shall be commenced within ___ (___) days from this date or approximately on (Date): ___ and shall be substantially completed within ___ (___) days or approximately on (Date): ___
Commencement of work shall be defined as (Briefly Describe Type Of Work Representing Commencement):   **SAMPLE**

PAYMENT: Owner agrees to pay Contractor a total cash price of $___ down payment (if any) $___ **SAMPLE**

Should contract call for a salesman's commission to be paid out of contract price, said payment shall be made by dispersing party to the Contractor.

Payment schedule as follows:

	WHEN		WHEN
$ __**SAMPLE**__		$ _____	
$ _____	**SAMPLE**	$ _____	**SAMPLE**
$ _____		$ _____ **SAMPLE**	

Upon satisfactory payment being made for any portion of the work performed, the Contractor shall, prior to any further payment being made furnish to the person contracting for the home improvement or swimming pool a full and unconditional release from any claim or Mechanic's Lien, for that portion of the work for which payment has been made.

ALLOWANCES: The following items or specific prices as indicated are included in the contract price as allowances. The contract price shall be adjusted upward / downward based upon actual amounts rather than estimated amounts herein ___ **SAMPLE**

TERMS AND CONDITIONS: The terms and conditions attached are expressly incorporated into this Agreement.

You, the buyer, may cancel this transaction at any time prior to midnight of the third business day after the date of this transaction. See the attached Notice of Cancellation form for an explanation of this right. Cancellation by the buyer after the right to rescind has passed, shall be deemed a material breach of this agreement and entitles Contractor to damages.

Firm Name: _____   Date: _____
(Contractor's Firm Name, If Any)
Name and State Registration Number of any salesman who solicited or negotiated this contract.

By: __**SAMPLE**__   Signature: _____
(Contractor Or Agent Sign Here)

Owner: **X** _____   Date: _____

Owner: **X** _____   Date: _____
(If More Than One Owner, Please Sign Here)

TERMS AND CONDITIONS

1. CHANGES IN THE WORK. Should the Owner, construction lender, or any public body or inspector direct any modification or addition to the work covered by this contract, the contract price shall be adjusted accordingly.

Modification or addition to the work shall be executed only when both the Owner and the Contractor have signed a Contract Change Order. The change in the contract price caused by such contract Change Order shall be as agreed to in writing, or if the parties are not in agreement as to the change in contract price, the Contractor's actual cost of all labor, equipment, subcontracts and materials, plus a contractor's fee of **(FILL IN PERCENTAGE) %** shall be the change in contract price. The Change Order may also increase the time within which the contract is to be completed.

No Extra or Change Order work shall be required to be performed without prior written authorization of the person contracting for the construction of the home improvement or swimming pool. Any Change Order forms for changes or Extra Work shall be incorporated in, and become a part of the contract.

2. RESPONSIBILITIES OF THE PARTIES. Contractor shall promptly notify the Owner of (a) subsurface or latent physical conditions at the site differing materially from those indicated in this contract, or (b) unknown physical conditions differing materially from those ordinarily encountered and generally recognized as inherent in work of the character provided for in this contract. Owner as added work shall pay for any expense incurred due to such conditions.

The Owner is responsible to supply water, gas, sewer and electrical utilities unless otherwise agreed to in writing. Electricity and water to the site is necessary.

Owner agrees to allow and provide Contractor and his equipment access to the property and provide toilet facilities.

The Owner is responsible for having sufficient funds to comply with this agreement. This is a cash transaction unless otherwise specified.

The Owner is responsible to remove or protect any personal property and Contractor is not responsible for it or for any carpets, drapes, furniture, driveways, lawns, shrubs, etc.

The Owner will point out and warrant the property lines to contractor.

3. DELAYS. Contractor agrees to start and diligently pursue work through to completion, but shall not be responsible for delays for any of the following reasons: failure of the issuance of all necessary building permits within a reasonable length of time, funding of loans, disbursement of funds into funding control or escrow, acts of neglect or omission of Owner or Owner's employees or Owner's agent, acts of God, stormy or inclement weather, strikes, lockouts, boycotts, or other labor union activities, Extra Work ordered by Owner, acts of public enemy, riots or civil commotion, inability to secure material through regular recognized channels, imposition of government priority or allocation of materials, failure of Owner to make payments when due, or delays caused by inspection or changes ordered by the inspectors of authorized governmental bodies, or for acts of independent contractors, or holidays, or other causes beyond Contractor's reasonable control.

4. PLANS & SPECIFICATIONS. If plans and specifications are prepared for this job, they shall be attached to and become apart of the agreement.

5. SUBCONTRACTS. The Contractor may subcontract portions of this work to properly licensed and qualified subcontractors.

6. FEES, TAXES AND ASSESMENTS. Owner will pay for taxes and assessments of all descriptions. Contractor will obtain and pay for all required building permits, but Owner will pay assessments and charges required by public bodies and utilities for financing or repaying the cost of sewers, storm drains, water service, schools and school facilities, other utilities, hook-up charges and the like.

7. COMPLETION AND OCCUPANCY. Owner agrees to sign and record a Notice of Completion within five (5) days after the project is complete and ready for occupancy. If the project passes final inspection by the public body but Owner fails to record Notice of Completion, then Owner hereby appoints Contractor as Owner's agent to sign and record a Notice of Completion on behalf of Owner. This agency is irrevocable and is an agency coupled with an interest. In the event the Owner occupies the project or any part thereof before the Contractor has received all payment due under this contract, such occupancy shall constitute full and unqualified acceptance of all the Contractor's work by the Owner and the Owner agrees that such occupancy shall be a waiver of any and all claims against the contractor.

8. INSURANCE AND DEPOSITS. Owner will procure at Owner's expense and before the commencement of any work hereunder, fire insurance with course of construction, vandalism and malicious mischief clauses attached, such insurance to be a sum at least equal to the contract price with loss, if any, payable to any beneficiary under any deed of trust covering the project, such insurance shall also name the Contractor and any subcontractors as additional insured, and to include sufficient funds to protect Owner, Contractor, subcontractors and construction lender as their interests may appear. Should Owner fail to do so, Contractor may procure such insurance as agent for and at the expenses of Owner, but is not required to do so.

If the project is destroyed or damaged by accident, disaster or calamity, such as fire, storm, earthquake, flood, landslide, or by theft or vandalism, any work done by the Contractor in rebuilding or restoring the project shall be paid by the owner as extra work.

Contractor shall carry Worker's Compensation insurance for the protection of Contractor's employees during the progress of the work. Owner shall obtain and pay for insurance against injury to Owner's own employees and persons under Owner's direction and persons on the job site at Owner's invitation.

9. RIGHT TO STOP WORK: Contractor shall have the right to stop work if any payment shall not be made, when due, to Contractor under this agreement. Contractor may keep the job idle until all payments due are received. This remedy is in addition to any other right or remedy that the Contractor may have. Such failure to make payment when due, is a material breach of this agreement. Owner acknowledges that the additional costs for the delay in stopping and starting the project shall be treated as an extra and allow Contractor additional costs in accordance with paragraph one hereof.

10. CLEAN-UP. Contractor will remove from Owner's property debris and surplus material created by this operation and leave it in a neat and broom clean condition.

11. LIMITATIONS. No action of any character arising from or related to this contract, or the performance thereof shall be commenced by either party against the other more than two years after completion of the project or cessation of work under this contract.

12. COMPLIANCE WITH LAWS. In connection with the performance by Contractor, pursuant so this agreement, Contractor shall obtain and pay for all permits and comply with all federal, state, county and local laws, ordinances and regulations.

13. ATTORNEY FEES. In the event there is any litigation or arbitration arising out of this agreement, the prevailing party shall be entitled to its reasonable attorney fees and costs.

14. PAYMENT. Upon satisfactory payment being made for any portion of the work performed, the Contractor shall, prior to any further payment being made, furnish to the persons contracting for the improvement, a full and unconditional release from any claim or Mechanic's Lien for that portion of the work for which payment has been made.

15. ASBESTOS AND HAZARDOUS WASTE. Unless the contract specifically calls for the removal disturbance, or transportation of asbestos or other hazardous substances, the parties acknowledge that such work requires special procedure, precautions, and/or licenses. Therefore, unless the contract specifically calls for same, if Contractor encounters such substances, Contractor shall immediately stop work and allow the Owner to obtain duly qualified asbestos and/or hazardous material contractor to perform the work or the Contractor may perform the work at contractor's option. Said work will be treated as an extra under the contract

Sample Architectural Drawings

The following drawings illustrate what you can expect to receive from an architect, home designer, or your contractor. The basic drawings are called a plan view and an elevation. A plan view is the floor plan. It is like looking at your house from the top without the roof on (page 143). You can see how all the walls and rooms are laid out. An elevation view is your house as it is seen from the outside, looking at it from approximately ground level (page 144). You will also likely have drawings of details that tell the builder how specific portions of your remodel project should be built (page 145). The site plan shows how your house is orientated on your lot (page 146).

Resources

A wealth of information is available from a variety of sources if you would like to learn more about remodeling, moving, or home financing. I have used many of these resources for my own projects as well as for general reference. They are categorized into three groups. First are the remodeling resources. They have excellent information on costs to remodel, do-it-yourself remodeling, products, and finding a qualified contractor. The home purchase and selling resources provide an abundant source of information on selling your home, buying your home, and moving. The home plans and design resources offer ideas on floor plans and decorating.

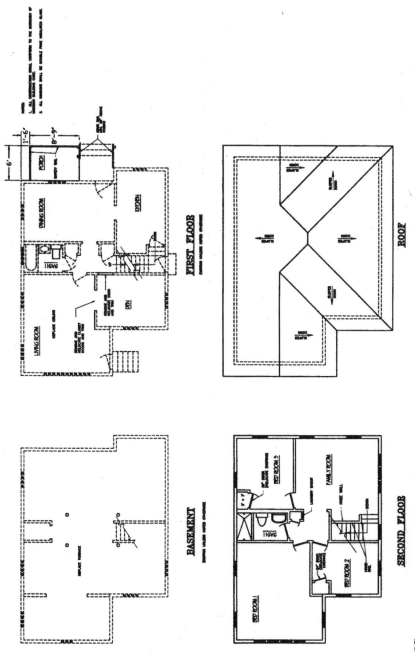

PLANS (Reprinted with permission. Copyright © 2004 Brian A. Falconer.)

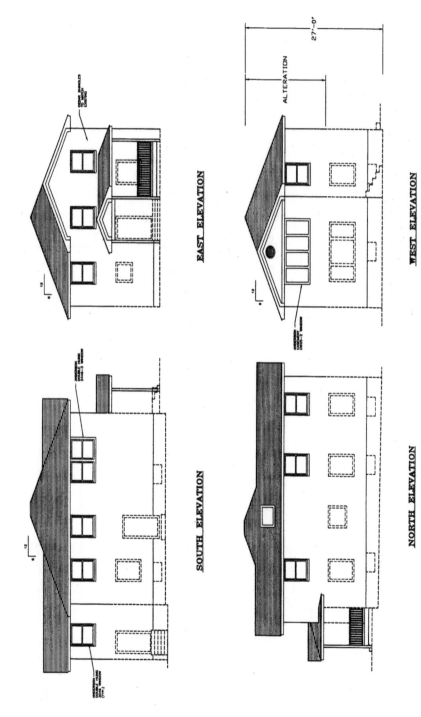

EAST ELEVATION

WEST ELEVATION

SOUTH ELEVATION

NORTH ELEVATION

(Reprinted with permission. Copyright © 2004 Brian A. Falconer.)

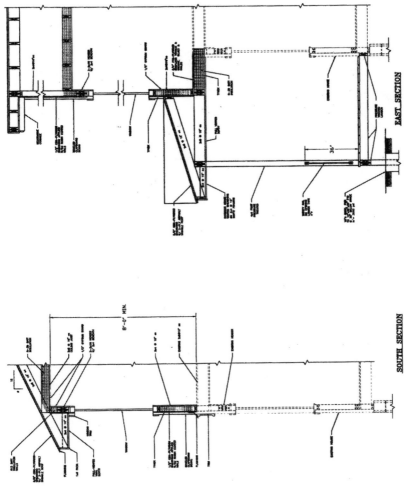

SECTIONS (Reprinted with permission. Copyright © 2004 Brian A. Falconer.)

SITE PLAN (Reprinted with permission. Copyright © 2004 Brian A. Falconer.)

SECOND STORY ALTERATION
PORCH ADDITION

Remodeling Resources

www.improvenet.com
Contractor referral service
How-to tips

www.remodelingmagazine.com
Online magazine for contractors
Product and building information

www.remodeltoday.com
Web site for the National Association of the
    Remodeling Industry
Information on products and how to work with contractors

www.servicemagic.com
Contractor referral service
How-to tips

www.remodelormove.com
Web site for this book
Workbooks and additional resources for remodeling

www.abcaforms.com
Provider of construction software and forms
Construction and remodeling documents

Home Purchase, Selling, Financing, and Moving Resources

www.realtor.com
Access to home-for-sale listings
General home reference and moving tips

www.homeseekers.com
Information on new and resale homes
Demographics on neighborhoods and cities

www.bankrate.com
A variety of financial and loan information

www.city-data.com
www.bestplaces.net
Statistics such as population, median income, and crime
   rate for cities throughout the United States

www.imove.com
Resource for do-it-yourself moving and using a professional
   moving company

www.homevaluehunt.com
www.homeradar.com
Lists of recent home sales near an address that you enter

www.moneycentral.msn.com
Articles and resources to manage money and finances

www.remodelormove.com
Workbooks and additional resources for home selling and
   moving

Home Plans and Design Resources

www.bhg.com
Web site for *Better Homes and Gardens* Magazine
A large variety of information for homeowners

www.remodelormove.com
Resources and services to create drawings for your remodel
   project

www.smartdraw.com
Software for drawing home floor plans

Notes

1. Steve Berges, *101 Cost Effective Ways to Increase the Value of Your Home* (Chicago, Dearborn Trade, 2004).

2. *Existing-Home Sales Hit Record* in May Press Release (National Association of Realtors, 2004).

3. *NARI Announces National Winners of the 2004 Achievement Awards* Press Release (National Association of the Remodeling Industry, 2004).

Glossary

A

ABS A type of black plastic pipe commonly used for waste water lines.

Acceptance Agreement to a contract.

Accrued interest Interest owed but not yet paid.

Acre A measure of land equal to 43,560 square feet.

Adjustable rate mortgage loans (ARM) Loans with interest rates that are adjusted periodically based on changes in a preselected index.

Adjusted basis The original cost of a property plus the cost of any improvements such as an addition.

Air ducts Pipes that carry warm or cold air into rooms and back to the furnace or air-conditioning system. Part of the HVAC system (heating, ventilation, air conditioning).

Aggregate Gravel or pebbles that are added to concrete and sometimes left exposed on the surface of the concrete for decoration.

Amortization Payment of a loan with regular installments calculated to pay off the loan at the end of a fixed period of time.

Amperage or amps A unit of electrical current or volume. Most homes have an electrical service that provides between 100 and 300 amps.

Anchor bolts Bolts that are set in the concrete foundation and used to attach the framing of the house to the foundation.

Annual percentage rate (APR) A percentage that results from an equation considering the amount financed, the finance charges, and the term of the loan. It is not necessarily the loan's interest rate. APR was created to help consumers more easily compare loans.

Application An initial statement of personal and financial information, which is required to approve your loan.

Application fee Fee charged by a lender to cover the initial costs of processing a loan application.

Appraisal An estimate of a property's current market value.

APR See annual percentage rate.

Appreciation An increase in the value or worth of a property; opposite of depreciation.

ARM See adjustable rate mortgage loans.

Asbestos Formerly a common building material because of its fire resistance. For more information see the EPA Web site or look for asbestos removal firms in the phone book.

Asking price The price placed on a property for sale. Also known as the list price.

Asphalt A black waterproof material used for roofing and pavement.

Assessor A local government official who determines the value of the property for taxation purposes.

B

Backfill Dirt or gravel used to fill in against a wall or foundation.

Backflow A reverse flow of water into the water supply pipes, caused by negative pressure in the pipes.

Balloon mortgage A mortgage that has regular payments for a period of time, typically three to seven years, and a large single payment at the end of the period to pay off the entire balance.

Baseboard A decorative board placed above the floor against walls and partitions to hide gaps.

Batt Insulation in the form of a blanket, rather than loose filling. Commonly used for insulation in new construction of walls, floors, and ceilings.

Beam A horizontal framing member designed to carry a load between two walls or other supports.

Bearing wall A wall that supports the ceiling, roof, or second story.

Bill of sale A written document given to pass the title of a property between people.

Blueprint A method of copying architectural drawings. Can also refer to the drawings themselves.

Board and batten A style of exterior siding where seams between wide vertical boards are covered by narrow vertical boards.

Bona fide In good faith.

Brokerage For a commission or fee a company that brings together parties interested in buying, selling, exchanging, or leasing real property.

Building paper Heavy paper used on exterior walls and under tile in showers to protect the framing from water and moisture.

Bump out A small addition that does not necessarily have a foundation because it is cantilevered or suspended.

Buy-down A payment to a lender to reduce the interest rate of a loan.

C

Cantilever Any part of a structure that projects beyond its main support and is balanced on it.

Capital gains The profit on the sale of a capital asset, such as stock or real estate.

Casement window A window with hinges on one of the sides that swings open like a door.

Cash-out Cash given to you when you get a new loan that is larger than the remaining balance of your current mortgage.

Casing Door and window framing.

Cast-iron pipe Heavy metal pipe used for plumbing drain and vents. Most new construction will use ABS plastic instead.

Caulking A flexible material used to seal a gap between two surfaces. Frequently used around windows, between a bathtub and tile, and to fill gaps in exterior siding.

Ceiling (financial) The maximum allowable interest rate of an adjustable rate mortgage.

CC&Rs See covenants, conditions, and restrictions.

Chair rail Molding on a wall at the height of a chair back.

Change order A document that modifies the terms, specifications, or price of a contract.

Circuit breaker A switchlike device that allows a maximum of current (amperes) into the wiring of a home. Above that current it will "break" or disconnect the wiring from the electrical service. The device can be reset by moving the switch level back to the original position.

Cistern A tank to catch and store rainwater.

Closing The conclusion of your real estate transaction; includes the delivery of the property title, signing of your legal documents, and the distribution of the funds necessary for the sale of your home.

Code Rules created and enforced by governments and trade organizations to set required building design and construction requirements.

Clear title A land title that doesn't have any liens (including a mortgage) against it.

Closing The conclusion of the sales transaction, when the seller transfers title to the buyer.

Closing costs Costs for services performed before your loan can be initiated. Examples include title fees, recording fees, appraisal fee, credit report fee, pest inspection, attorney's fees, and surveying fees.

Closing statement A detailed written summary of the financial settlement of a real estate transaction, showing all charges and credits made and all cash received and paid out.

COFI See cost-of-funds index.

Commission Money paid to a real estate agent or broker for negotiating a real estate or loan transaction.

Comparables Properties that are similar to a particular property used to compare and establish a value for that property.

Concrete block A hollow concrete "brick," often 8 inch x 8 inch x 16 inch in size, that is used to build walls.

Concrete board A panel made out of concrete and fiberglass used as a tile-backing material.

Conforming loan A mortgage loan that meets all requirements to be eligible for purchase by federal agencies.

Contingency A requirement in a contract stating that some or all of the terms of the contract will be void unless certain specified requirements are met.

Contractor A company licensed by the state or other government to perform construction and remodeling projects. There are various types of contractors:

General contractor A contractor that takes responsibility to organize and coordinate all the tradespeople necessary to complete a remodeling project.

Subcontractor A contractor that works for a general contractor.

Convertible ARMs A type of ARM loan with the option to convert to a fixed-rate loan during a given time period.

Conveyance The document used to effect a transfer, such as a deed, or mortgage.

Cost-of-funds index (COFI) An index of the weighted-average interest rate paid by savings institutions for sources of funds, usually by members of the 11th Federal Home Loan Bank District.

Covenants, conditions, and restrictions (CC&Rs) The restrictions governing the use of real estate, usually enforced by a homeowners' association and passed on to the new owners of a property.

Crawl space A shallow, unfinished space beneath the first floor of a house that has no basement.

Credit report A report detailing the credit history of a prospective borrower that's used to help determine borrower creditworthiness.

Credit insurance Insurance that will pay off a loan if a borrower dies or becomes disabled or in case of other events that are covered by the policy.

Credit limit The maximum amount that you can borrow under a home equity plan.

Credit report A summary of your loan activity and payment history; used to determine if you are "safe" to lend money to.

Credit score A calculation that provides a single number indicating your creditworthiness. It is a single number that summarizes your credit report.

Creditor A person or bank to whom a debt is owed.

D

Deed A legal document by which title to real property is transferred from one owner to another.

Debt-to-income ratio The measurement of debt payments to gross household income used to determine how much can be lent to a borrower.

Default Failure of one or both parties to meet legal obligations in a contract.

Delinquency Failure to make payments as agreed.

Depreciation A loss in value of an asset, such as a home or a car.

Disclosure The making known of a fact that may not be obvious about a home that is for sale.

Discount points A fee paid to lender at the time that you get your loan to reduce the interest rate or otherwise change the terms of the loan. Also called points.

Double-hung window A window with two vertically sliding sections; very common in older homes and new homes of certain styles.

Drywall The most common material used to finish interior walls. Sheets are screwed or nailed to framing, dents and depressions are filled, and a texture applied.

Dormer A projecting section of framing from a sloping roof, usually containing a window or vent.

Dry rot A fungus that slowly destroys wood.

E

Earthquake strap A metal strap used to secure gas hot water heaters to the framing or foundation of a house.

Easement An agreement that allows a party to use a portion of your property for a specific purpose.

Equity The difference between the current market value of a home and the total mortgage owed on the same home.

Escrow A transaction in which a third party acts as the agent for seller and buyer or for borrower and lender in handling legal documents and disbursement of funds.

F

Fannie Mae A company created by the U.S. government to "buy" home loans from mortgage companies, allowing more loans to be made.

Federal Housing Administration A federal government department that insures mortgage loans made by approved lenders in accordance with FHA regulations.

FHA See Federal Housing Administration.

First mortgage A mortgage that will be paid first from proceeds from the sale of the home if a homeowner defaults on payments.

Fixed-price contract A contract with a set price for the work.

Fixed rate An interest that doesn't change for the term of the loan.

Footing A widened base of a foundation wall.

For sale by owner A home that is being sold without a real estate agent.

Framing The structural wood and/or metal elements of most homes.

FSBO See for sale by owner.

Fuse A device found in older homes designed to prevent overloads in electrical lines.

G
GFI or GFCI See ground fault current interrupter.

Good-faith estimate Written estimate of the settlement costs the borrower will likely have to pay at closing.

Grace period The period of time during which a loan payment may be made after its due date without incurring a late penalty.

Gross income Your total income before taxes or expenses are deducted.

Ground fault current interrupter An electrical device built into normal electrical outlets that can prevent injury from contact with electrical appliances. Required in new homes in bathrooms, kitchens, garages, and outdoors.

H

Hardwood A very durable type of wood from broad-leaved trees, such as oak or maple, commonly used for floors, cabinets, and furniture.

I

Impound account An account held by a lender to which the borrower pays monthly installments, collected as part of the monthly mortgage payment, for annual expenses such as taxes and insurance. The lender disburses impound account funds on behalf of the borrower when the expenses are due.

Interest The charge paid for borrowing money, calculated as a percentage of the remaining balance of the amount borrowed.

Interest rate The annual rate of interest on the loan, expressed as a percentage of 100.

J

Jamb An exposed frame that retains a window or door.

Joists One of a number of small, parallel beams for supporting a floor or ceiling.

L

Lath One of a number of thin narrow strips of wood nailed to rafters, ceiling joists, wall studs, and so on to make a surface for applying plaster.

Lien A legal claim by one person on the property of another for security for payment of a debt.

Load-bearing wall A wall that is supporting other parts of a home such as the roof or a second story.

Loan origination fee A fee charged by a lender to cover the administrative costs of processing a loan.

Loan-to-value ratio The percentage of the loan amount compared to the appraised value (or the sales price, whichever is less) of the property.

LTV See loan-to-value ratio.

M

Market value The price that a buyer can buy a home from a seller.

MLS See multiple listing service.

Moisture barrier Treated paper or plastic that prevents water or water vapor from coming in contact with the protected material.

Molding A strip of decorative material having a plain or curved narrow surface. These strips are often used to hide gaps at wall junctures.

Mortgage A legal document by which real property is pledged as security for the repayment of a loan.

Mortgage insurance Insurance to protect the lender in case you default on your loan. Also called private mortgage insurance (PMI).

Multiple Listing Service A directory where member real estate agents can advertise the homes that they have listed.

N

Negative amortization A loan payment schedule that causes the principal balance of a loan to increase instead of decrease.

Note A legal document obligating a borrower to repay a loan at a stated interest rate during a specified period of time.

O

Open listing A listing under which the principal (owner) reserves the right to list his property with other brokers.

P

Per diem interest Interest calculated per day.

Permit Authorization from a government organization to remodel.

Pier block A concrete block used to support foundation members such as posts, beams, girders, and joists.

Pitch The angle of the slope of a roof as expressed as the ratio of the rise, in feet, to the span, in feet.

PITI Principal, interest, taxes, and insurance.

Plywood A sheet of wood manufactured by gluing several sheets together.

Points (or discount points) A fee paid to the lender at the time that you get your loan to reduce the interest rate or otherwise change the terms of the loan.

Power of attorney A legal document authorizing one person to act on behalf of another.

PMI See private mortgage insurance.

Prefabrication Construction of components, such as walls, trusses, or doors, before delivery to a building site.

Prepayment penalty Fee charged by a lender for a loan paid off in advance of the contractual due date.

Prequalification The process of determining how much money a prospective homebuyer will be eligible to borrow prior to applying for a loan.

Principal The amount of debt, not counting interest, left on a loan.

Private mortgage insurance (PMI) Insurance to protect the lender in case you default on your loan.

Prorate To divide or distribute equally.

R

Rent back Staying for a few months in your home after you sell it by paying the new owners a monthly payment.

Resilient flooring A durable floor such as vinyl or linoleum.

Right to rescission Under the provisions of the Truth-in-Lending Act, the borrower's right, on certain kinds of loans, to cancel the loan within three days of signing.

Roof sheathing Sheets, usually of plywood, nailed to the top edges of trusses or rafters to tie the roof together and support the roofing material.

Rough flooring Materials used to form an unfinished floor. Also known as the subfloor.

S

Sash The movable part of a window; the frame in which panes of glass are set in a window or door.

Septic tank A sewage-settling tank in which part of the sewage is converted into sludge before the remaining waste is discharged by gravity into the ground.

Setback The distance a building must be set back from the property lines in accordance with local zoning ordinances or deed restrictions.

Shake A wood shingle used for roofing.

Sheathing Exterior-grade boards used to enclose the exterior of a home.

Slab A type of foundation with a concrete floor placed directly on the soil.

Soffit The finished underside of the eaves.

Subfloor Usually, plywood sheets that are nailed directly to the floor joists and that receive the finish flooring.

Sump A pit in the basement in which water collects to be removed with a pump.

T

Tear down A house that will be substantially removed from a lot and a new one rebuilt in its place.

Time and material contract An agreement with contractors where they are paid both an hourly rate and for the cost of the materials they buy to complete the work.

Treated lumber A wood product impregnated with chemicals to reduce damage from wood rot or insects. Often used for the portions of a structure that are likely to be in ongoing contact with soil and water.

U

Underlayment A layer of wood or other material designed to support a surface material; used under vinyl and other types of flooring as well as beneath shingles or other roofing material.

Underwriting The process of verifying data and approving a loan.

W

Wainscoting The lining on the lower three or four feet of an interior wall, consisting of paneling different from the rest of the wall.

Wall sheathing Sheets of plywood, nailed to the outside face of studs as a base for exterior siding.

Z

Zoning The city or county regulations that control the use and character of property.

Habitat for Humanity

NET PROFITS FROM THE sale of this book are donated to a local affiliate of Habitat for Humanity or other charitable organizations that promote homeownership.

Following are excerpts from the United States Department of Housing and Urban Development Publication, "Making Homeownership a Reality: Survey of Habitat for Humanity International Homeowners and Affiliates."

Habitat for Humanity International has produced more than 60,000 homes since its founding in 1976. As one of the most successful homeownership programs for low-income families, its housing production volume easily puts it in the ranks of the Nation's top 20 homebuilders. Making Homeownership a Reality: Survey of Habitat for Humanity International Homeowners and Affiliates, sponsored by the HUD Office of Policy Development and Research, presents key findings from a survey of Habitat homeowners.

Habitat and Affordability
Habitat primarily serves low- and very low-income families. Only 20 percent of homeowners believe that

they would have bought a home without Habitat's assistance. About 43 percent of households earned less than one-half of the median household income, while 34 percent earned between 50 and 80 percent of the median income.

Homeownership Outlook
According to the survey, the most common benefits of homeownership were the pride and increased stability that families received from feeling secure about their homes. Most homeowners planned to continue living in their homes and eventually pass them on to their children.

If you would like to learn more about Habitat for Humanity, visit www.habitat.org. You can contribute online to the central organization or one of 1,600 local affiliates, one of which is in your community. Donate money or your time. It will make you feel great.

About the Author

DAN FRITSCHEN, author of *Remodel or Move? Make the Right Decision,* has ten years' experience as a professional consultant and workshop speaker and thirty years' experience with real estate and remodeling. He has found that most people think remodeling costs more than it really does and that moving is free.

His own experience with moving and remodeling dates back to his childhood. His mother was a real estate agent and has owned and managed rental properties. His parents were always adding a window, moving a wall, building a deck, or painting a rental unit, so he and his four brothers and sisters pitched in to help from an early age.

Born near Washington, D.C., Dan moved to Michigan, where he attended middle and high school. After attending the University of Kansas, where he studied mechanical engineering and business, he moved to California to race sailboats, a longtime interest. He now lives in Silicon Valley in Northern California with his wife and two daughters.

He has owned six homes, all of them remodeled a little or a lot, and been involved in more than twenty remodeling projects, working on the design, managing the project, or actually doing the work. He recently helped a homeowner understand the total costs of moving as well as that there are many ways to remodel a home, some expensive and some inexpensive. The homeowner, armed with this information, made an informed remodel-or-move decision. The success of that project inspired this book.

Do You Need Remodel-or-Move Advice?

Two hours spent with a move-or-remodel consultant saved me more than $80,000 on my home remodeling project. I originally got a quote for an addition to my home for $160,000 from a well-respected local contractor. I had only done a little research and was unsure of what I should do so I made an appointment with a remodel-or-move consultant. The consultant discussed my goals with me and listed options of how I could proceed. After getting this advice, I hired different contractors to do each of three major phases of the project, reducing the total cost from $160,000 to $80,000. The advice from a remodel or move consultant was the best investment I have ever made.

—Saeed Taghipour, 2004

Would you like to get an expert and unbiased opinion about your move-or-remodel decision from someone you trust?

- Unbiased. We're not trying to sell you a house or get you to remodel.

- Fast and easy. Phone consultations can be conducted at a convenient time for you—morning, day, or night.

- Free follow-up questions via e-mail.

- Fully trained consultants to help you make the best decision.

Call 888-825-4169 or
e-mail advice@remodelormove.com
to schedule an appointment.

Ordering Information

Remodel or Move? Make the Right Decision $15.95
   Shipping and handling in the United States $ 4.00
   Add $1.00 for each additional copy.

Remodel or Move? Organizer and Workbook $23.95
   Shipping and handling in the United States $ 7.00
   Add $1.00 for each additional copy.

Mail orders:
ABCD Publishing LLC
1030 E. El Camino Real, Suite 150
Sunnyvale, CA 94087

Telephone orders:
888-825-4169

Fax orders:
866-620-9912

E-mail orders:
orders@remodelormove.com

Web orders:
www.remodelormove.com

Do you have any comments regarding *Remodel or Move? Make the Right Decision*? Is there a topic that should be included or other information that you need? Please let us know so we can continue to offer the only comprehensive guide to making the right decision.

YES, I want to make the right decision and save money.

Remodel or Move? Make the Right Decision

$15.95 x _____ copies = $ _____

Remodel or Move? Organizer and Workbook

$23.95 x _____ copies = $ _____

Shipping/handling $ _____

CA residents: add sales tax $ _____

TOTAL $ _____

❑ Check or money order enclosed (payable to ABCD Publishing LLC)

❑ Charge my ❑ VISA   ❑ MasterCard

Card number _____ Exp. Date _____

Signature _____

Name _____

Address _____

City _____ State _____ Zip_____

Daytime phone number _____ Fax _____

E-mail _____

❑ Please send specific information regarding remodel or move costs for my state, including tax information.

❑ Please send the "Remodel or Move" newsletter containing tips on how to make the decision and how to save money if I remodel or if I move.

Index

H
Habitat for Humanity, 173–174
heating, 53
heating systems, 29, 113–114
home equity loans, 117

I
ideas for remodeling, finding, 106
improvements
    for selling your home, 26–29
    types matrix, 67
    to your new home, 35
inconveniences of remodeling,
    8–10
inspections, 31–32, 35–36
interior improvements, 27–29
interior surfaces, 53
investment issues, 20–22, 48–49
invoices, 107

K
kitchen fixtures, 54–56

L
landscaping, 26
learning about remodeling, 127
liens, 126–127
lighting fixtures, 29
list of wants, 107–108
loan options, 117–118
location factors, 22–23

M
managing remodeling yourself,
    42–43
marketing your home, 30, 97–98
MLS (Multiple Listing Service), 30
moving
    costs, 2, 24
        buying your new home, 34–37
        getting ready, 26–29
        inspections, 31–32
        moving, 34
        real estate agent's commis-
            sion, 30–31

reasons for
    better schools, 7–8
    checklist for, 14–15
    expense of remodeling, 13–14
    floor plan/inability to
        remodel, 11–12
    houses in neighborhood, 14
    inconvenience of remodel-
        ing, 8–10
    length of commute, 8
    neighborhood, 10–11
    size of family, 5–7
    yard problems, 12–13
Multiple Listing Service (MLS), 30

N
necessary versus not recom-
    mended projects, 68
needs, changes in family, 5–7
negotiating, best deals for selling
    your home, 31
negotiating the sale, 98
neighborhoods, 10–11, 14, 19
net worth, 3

O
offers, accepting, 102, 135

P
packing/moving, 34, 102–104
paybacks of remodeling, 47–49
paying for a remodel, 117–118,
    120–121
permits for remodeling, 44, 45,
    46, 49–50, 107, 112
per-square-foot costs, 129–130
plumbing, 29, 52–53, 113
price setting, for selling your
    home, 30, 100–102
project management, 41–44, 111
property taxes, 36–37
purchase orders, 107

Q
quality-of-life issues, 3–4, 19, 22
quick-remodel decisions, 67